CURTIS'S
BOTANICAL MAGAZINE

15-Year Index

1984-1998

Volumes 1-15 of Series 6
Volumes 185-199 of the Whole Work

compiled
by
Pat Halliday and Brian Mathew

incorporating

The Kew Magazine

First Published 2002

ISBN 1 84246 042 0

Printed in Great Britain
by
Culver Limited

CURTIS'S BOTANICAL MAGAZINE

(including The Kew Magazine)
15-Year Index to Series 6, Volumes 1-15 (1984-1998)
Volumes 185-199 of the Whole Work
d = line drawing, m = map, p = portrait, pl = colour plate on facing page
For list of plates only, listed in A-Z sequence of species, see page 55

Abarema idiopoda 14(4): 198-202, 198pl, 210d

Abeliophyllum distichum 15(2): 141-146, 145d

Abies spectabilis 13(2): 100d, 101, 102d

Abutilon darwinii 1(2): Cover pl

Acacia 6(1):6; 14(4): 176, 177, 179, 181, 187, 243-244

  aulacocarpa 1(3): 126d

  chamelensis 10(2):66-69, 68pl, 68d

  heterophylla 13(4): 215-216

  karroo 14(4): 192

  melanoxylon 14(4): 192

  nilotica 14(4): 192

  pataczekii 6(1):6-8, 6pl, 7d

  tortilis 14(4): 192

  sect. *Filicinae* 10(2):67; 13(4): 189

Acanthophoenix 13(4): 186-187

Acanthostachys strobilacea 8(2): 67-69, 68pl, 68d

Acanthus dioscoridis 7(4):161-165, 162pl, 163d

  f. *boissieri* 7(4):164

  subsp. *brevicaulis* 7(4):164

  subsp. *longistylis* 7(4):164

  var. *brevicaulis* 7(4):164

  var. *grandiflorus* 7(4):164

  var. *straussii* 7(4):164

Acer caesium 13(2): 101

Aceras caprina 3(1): 31

  var. *calcarata* 3(1): 31

  hircina 11(1): 26

  hircina var. *caprina* 3(1): 31

Acorus calamus 13(2): 98

Acrostichum crinitum 8(3):122

Actinotinus sinensis 3(4): 183

Adansonia distribution 13(2): 96

  economic importance 13(2): 96

  fruit-animal interactions 13(2): 90

  pollination 13(2): 96

  taxonomy 13(2): 96

Adansonia 13(2): 86-95

  digitata 13(2): 86-90

  gibbosa 13(2): 87

  grandidieri 13(2): 88d, 89-95

  gregorii 13(2): 87

  madagascariensis 13(2): 97, 89

  perrieri 13(2): 87, 89-90, 92

  rubrostipa 13(2): 87, 88d, 90-91, 94

  suarezensis 13(2): 88d, 89-92, 95

  za 13(2): 87, 88d, 90, 92, 94

    var. *bozy* 13(2): 89

    var. *za* 13(2): 89

  subg. *Adansonia* 13(2): 89

  subg. *Brevitubae* 13(2): 89-90

  subg. *Longitubae* 13(2): 87, 89

Adelges 3(3): 121

Adiantum capillus-veneris 8(3):149d

  'Cornubiense' 8(3):149d

  'Depauperatum' 8(3):149d

  'Halliston' 8(3):149d

  'Lunulatum' 8(3):149d

  'Nobile' 8(3):149d

  'Pictum' 8(3):149d

  'Ramosum' 8(3):149d

  'Undulatum' 8(3):149d

Aechmea abbreviata 1(1): 6-8, 6pl, 7d; 2(1): 198

  chantinii 1(1): 6

  fasciata 1(1): 6

  streptocalycoides 2(1): 198-201, 198pl, 200d

Aegilops speltoides 2(4): 367
  tauschii 2(4): 367
Aenictophyton 11(4): 172
Aerangis 4(2):69; 13(3): 146-152
  alcicornis 13(3): 147-148
  arachnopus 11(2): 55-59, 56pl, 58d
  batesii 11(2): 58
  biloba 11(2): 55
  biloboides 11(2): 58
  brachycarpa 11(2): 55; 13(3): 146-147
  carnea 13(3): 149
  distincta 13(3): 146-152, 146pl, 151d
  'Kew Elegance' 13(3): 149
  erythurum 15(1): 24
  flabellifolia 11(2): 55
  fuerteventurae 15(1): 75
  kotschyana 13(3): 146; 15(1): 21, 22, 23
  kotschyana sensu Schelpe 15(1): 24
  mystacidii 11(2): 55
  schliebenii 15(1): 24
  spiculata 4(2):69-74, 70pl, 72d
  splendida 13(3): 147, 149
  stelligera 13(3): 149
  verdickii 15(1): 20-26, 20pl, 25d
Aeschynanthus, with notes on two
  unusual species 8(1):15-23
Aeschynanthus angustifolius 8(1):21-23,
  22pl, 23d
  chrysanthus 8(1):17-21,18pl, 20d
  stenophyllus 8(1):22
  tetraquetrus 8(1): 22
Aesculus indica 13(2): 101
Aframomum laxiflorum 3(2): 84d, 85
African Violet (*Saintpaulia*): 1(2): 92-93;
  3(2): 86-87; 15(1): 49-62, 62-67
Agapetes moorei 12(4): 241
  saxicola 1(2): 77
Ailanthus flavescens 15(3): 189
Ainsworth, Jean 15(2): 137-140, 154
Ainsworth, John 15(2): 137-140, 154
Alabama canebrake pitcher 15(2): 140
Alantsilodendron 14(4): 232
Albizia 14(4): 176, 179, 181, 199, 234,
  235, 242
  lebbeck 13(4): 219
Albuca abyssinica 14(1): 32
  pendula 14(1): 31-34, 32pl, 33d
  subg. *Albuca* 14(1): 32
  subg. *Eu-albuca* 14(1): 32
  subg. *Falconera* 14(1): 32
  subg. *Pallastema* 14(1): 32
Alerce Andino National Park Chile
  4(3):127-140
Alexa 14(4): 189
Alexander, Crynan 1(1): 2, 3-6; 1(3):
  119-122
Alliaceae 11(1): 3
Allium, How many species are known?
  9(4):180-182
Allium sect. *Briseis* 4(4):195, 196
  sect. *Megaloprasum* 4(2):95
  sect. *Schoenoprasum* 11(1): 3
  subg. *Rhizirideum* 11(1): 3
  elatum 4(2):94, 95
  giganteum 4(2):91-96, 93d
  macleanii 4(2):94, 95
  oleraceum 11(1): 24
  paradoxum 4(4):195, 196
  var. *normale* 4(4):194-199, 197d
  var. *paradoxum* 4(4):196, 197d,198
  pendulinum 4(4):198
  schmitzii 11(1): 3-5, 4pl, 5d
  schoenoprasum 11(1): 3
  thunbergii 6(2):63
  triquetrum 4(4):195, 198
  ursinum 4(4):198
  virgunculae 6(2): 61-64, 62pl, 62d
Alocasia cuprea 14(2): 82
  guttata 14(2): 83-84
  heterophylla 14(2): 83
  longiloba 14(2): 82
  melo 14(2): 82-86, 82pl, 85d
  peltata 14(2): 82
  robusta 14(2): 82
  sanderiana 14(2): 82
  scabriuscula 14(2): 83-84

villeneuvei 14(2): 83
zebrina 14(2): 82
Aloe deltoideodonta 3(2): 58
  rauhii 3(2): 57-61, 58pl, 60d
Alpine Ecology 1(1): 30-35
Alstroemeria 14(1): 7-12
Alstroemeriaceae 14(1): 7
Altai Mountains 14(1): 35-55
Alves, Ruy J.V. 13(1): 33-35
Amaracus amanus 11(3): 114
  dictamnus 11(3): 114
  rotundifolius 11(3): 114
  tomentosus 11(3): 114
Amaryllidaceae 11(1): 9; 14(3): 142
  tribe *Haemantheae* 11(1): 9
Amaryllis 6(3):95
  aurea 6(3):101
  chilensis 6(3):96
  flammea 6(3):101
  flava 6(3): 96
  peruviana 6(3): 101
  tubiflora 6(3):101
Amberboa 13(1): 4
  sect. *Stephanochilus* 13(1): 4
Amburana cearensis 2(1): 221
American basket-flower 13(1): 3
American Blackberry 5(1):32
Amesia royleana 10(1): 20
Amherstia 14(4): 181
Ammocharis sp. 11(1): 10
Amorpha 14(4): 182
Amphisiphon stylosa 6(2):58-61, 58pl,
  60d; 7(3):125, 126
Amydrium 12(2): 85-89
  hainanense 12(2): 86
  magnificum 12(2): 86-87
  zippleianum 12(2): 85-89, 86pl, 88d
Anacamptis pyramidalis 11(1): 24
Anadenanthera colubrina
  var. *cebil* 2(1): 220d
Anadendrum 12(2): 85
Anagyris 14(4): 182
Anaphalis spp. 13(2): 98

Andira 14(4): 187, 242-244
Andrews, Milton 10(4):170d
Andrews, Susyn 1(3): 99, 100-103, 188-
  189; 3(3): 125-140; 4(3):123-127;
  5(2):76-82; 7(4):179-182; 8(2):71-78;
  9(1):30-35; 12(4): 236-244
Androcymbium species of the Canary Is.
  15(1): 67-77
Androcymbium 15(1): 67
  gramineum 15(1): 70-72
    subsp. *psammophilum* 15(1): 75
  hierrense 15(1): 68-77
    subsp. *hierrense* 15(1): 75
    subsp. *macrospermum* 15(1): 75
  psammophilum 15(1): 68-77
  punctatum 15(1): 70-71
  sect. *Erythrostictus* 15(1): 68
Androsiphon capense 7(3):124-128, 127d,
  124pl
Anemone obtusiloba 12(4): 243
'*Anemone coum*' 3(2): 95
  '*mirabile*' 3(2): 95
Angel, Rosemary 10(2):92, 10(3):135
Angiopteris 8(3):136
  elliptica 8(3):133-140, 134pl, 135d, 137d,
  139d
Angorkis arachnopus 11(2): 58
Angraecum alcicorne 13(3): 147
  angustum 15(1): 24
  arachnopus 11(2): 55
  batesii 11(2): 55, 58
  bicaudatum 15(4): 249
  biloboides 11(2): 58
  bilobum 11(2): 55
  eburneum 11(1): 36; 13(3): 162
  erythrurum 15(1): 24
  fuscatum 4(2): 73
  mystacidii 11(2): 55
  sesqipedale 13(3): 161
  verdickii 15(1): 24
Anguria warscewiczii 11(3): 151
Anisogonium attenuatum 8(3):131
Annona squamosa 2(1): 221

3

Another *Daphne* from Crete 11(3): 117-122

Anquetilia laureola 4(4):176, 191

Ansellia, The genus 10(1):27-35

  africana 10(1):27-35, 30pl, 33d

    var. *africana* 10(1): 30

    var. *australis* 10(1): 30, 34

    var. *natalensis* 10(1): 33

    var. *nilotica* 10(1): 34

  confusa 10(1):34

  congoensis 10(1): 34

  gigantea 10(1): 33

    var. *nilotica* 10(1):30, 34

  humilis 10(1):34

  nilotica 10(1):34

Ant Houses 12(2): 89-90

Anthocercis viscosa 15(4): 257, 258d, 259d, 260pl, 260d, 261d

Anthurium warocqueanum 12(3): 127-130, 128d,128pl

  sect. *Cardiolonchium* 12(3): 127

Antirrhinum grosii 6(3):106-109,106pl, 107d

Aphyllorchis 1(2): 88-89

Araceae 12(2): 85-89, 12(3): 122-176; 13(1): 14-18; 14(2): 82

  Acolytes of the 12(3): 153-168

  Introduction to Family 12(3): 122-125

  Genera of *Araceae* Project 12(3): 125-126

  subfam. *Aroideae* 12(3): 135

  subfam. *Lasioideae* 12(3): 138

  subfam. *Monsteroideae* 12(2): 86

  subfam. *Pothoideae* 12(2): 86, 12(3): 138

  tribe *Anthureae* 12(3): 138

  tribe *Colocasieae* 12(3): 145

  tribe *Culcaseae* 12(3): 138

  tribe *Nephthytideae* 12(3): 138

  tribe *Zomicarpeae* 12(3): 135

Arachis longisepala 11(1): 34

Arachnis calcarata subsp.

  longisepala 9(2):57

  longisepala 9(2):55-59, 56pl, 58d

  x *maingayi* 9(2):55

Arachnites lutea var. *minor* 3(1): 20

Araucaria bidwillii: Andrew Petrie's Pine 8(4):175-185

Archer River, Cape York Peninsula 1(3): 122-132

Archidendron 14(4): 199

Arctostaphylos glauca 15(3): 164-168, 164pl, 167d

  var. *puberula* 15(3): 165

Arctotis 'Nicholas Hind' 9(4):160-164, 162pl, 163d

Arderne, H.M. 11(1): 38

Arenaria norvegica 2(3): 317d

Argania spinosa 9(2):76-85, 77d

Argan: multipurpose tree of Morocco 9(2):76-85

Argent, George 5(3):99-119; 8(2):63-71; 9(1):18-21; 12(2): 77-85

Ariocarpus agavoides 1(4): 184-185, 185d

  scapharostrus 1(4): 184, 184d

Arisaema 1(2): 51-54

  bottae 4(2):67-68, 68pl, 68d

  candidissimum 13(2): 108-111, 109d

  Rediscovery in the wild 13(2): 108-111

  consanguineum 1(2): 59-61, 60d, 60pl, 62

  costatum 1(2): 55-57, 55d, 56pl, 60

  curvatum 1(2): 57

  exappendiculatum 4(3):114-116, 114pl, 115d

  filiforme 2(4): 348-350, 348pl, 350d

  hypoglaucum 1(2): 59, 61

  kerrii 1(2): 59, 61

  leschenaultii 1(2): 61-63, 62d, 62pl

  papillosum 1(2): 62, 63

  sazensoo 3(3): 109-110

  serratum 3(3): 110

  sikokianum 3(3): 109-112, 110pl, 111d

    var. *serratum* 3(3): 110

  sugimotoi 1(2): 67

  tortuosum 1(2): 57-59, 58d, 58pl

  tripartitum 5(1):21

  triphyllum 1(2): 52, 54

  undulatifolium 1(2): 64-65, 64d, 64pl

ssp. *undulatifolium* 1(2): 64d
wightii 1(2): 57
yamatense 1(2): 64, 66-67
   var. *sugimotoi* 1(2): 67
   ssp. *sugimotoi* 1(2): 66-67, 66d,66pl
   ssp. *yamatense* 1(2): 66
Arisaron 7(1):17
Arisarum 7(1): 14-21
  proboscideum 7(1):18-19, 14pl, 18d
  simorrhinum 7(1):19, 18pl, 19d
  vulgare 7(1): 20-21, 20d, 20pl
Aristolochiaceae 11(2): 59
Aroid Conservation 12(3): 173-176
Aroids, Cultivation at Kew 12(3): 168-173
Arrojadoa 13(2): 70-78
  key to species of 13(2): 78
  two species of 13(2): 70-78
Arrojadoa bahiensis 13(2): 72-75, 72pl,
  74d
  cremnophila 13(2): 74
  dinae 13(2): 70, 73, 75
  penicillata 13(2): 75-78, 74d, 76pl
   var. *decumbens* 13(2): 77
   var. *spinosior* 13(2): 77
  rhodantha 13(2): 70-71, 76
Artillery Plant (*Pilea microphylla*) 1(1): 17
Artocarpus altilis 10(2): 85; 13(4): 222
Arum arisarum 7(1):14
  balearicum 5(2): 76
  corsicum 5(2): 76
  costatum 1(2): 57
  filiforme 2(4): 350
  pictum 5(2): 72-76, 72pl, 75d
  tortuosum 1(2): 59
Arundinaria simonii 8(4):189
  spp. 13(2): 103
Asarum species, Two Asian 11(2):
  59-65
Asarum, Two Japanese species of 9(2):
  67-75
Asarum caulescens 11(2): 59-63, 60pl, 62d
  caulescens var. *setchuenense* 11(2): 61
  franchetianum 11(2): 61

fudsinoi 9(2): 68-72, 68pl, 70d, 71d
hirsutisepalum 11(2): 63
kumageanum 11(2): 63
  yakusimense 11(2): 63-65, 62pl, 64d
  minamitanianum 9(2): 72-75, 72pl, 75d
Asclepias carnosa 7(1):5
Aspidium fadyenii 8(3):146
  hookeri 8(3): 146
  proliferum 8(3): 146
  tuberosum 8(3): 115
Asplenium attenuatum 8(3): 131
  decussatum 8(3): 131
  luzoniense 8(3): 131
  marinum 8(3): 154
  'Acutum' 8(3): 151d
  'Attenuatum' 8(3): 151d
  'Crenatum' 8(3): 151d
  'Cuneatum' 8(3): 151d
  'Exaltatum' 8(3): 151d
  'Incisum' 8(3): 151d
  'Minor' 8(3): 151d
  'Minus' 8(3): 151d
  'Multipinnatum' 8(3): 151d
  'Plumosum' 8(3): 151d
  'Ramosum' 8(3): 151d
  'Ramo-trapeziforme' 8(3): 151d
  mauritiensis 13(4): 233
  proliferum Lam. 8(3): 131
  proliferum Sw. 8(3): 146
  spinulosum 8(3): 131
Aster litoralis 4(3): 123
  oharai 4(3): 123
  quitensis 8(1): 3
  spathulifolius 4(3): 123
Astilbe, National Collections 15(2): 136
Astilbe davidii 15(2): 134pl(b/w)
Astragalus 14(4): 177, 180
Ateleia 14(4): 187, 192, 243
Atherurus tripartitus 5(1): 21
Athyrium accedens 8(3): 131
  filix-femina 8(3): 153d
  'Bracklin Falls' 8(3): 153d
  'Chelsea' 8(3): 153d

'Keswick' 8(3): 153d
'Oban' 8(3): 153d
ridleyi 8(3):131
Atkins, Sandy 11(4): 174-178; 13(1): 33-
35; 13(2): 79-81
Augusta attenuata 9(3): 130
glaucescens 9(3): 130
lanceolata 9(3): 130
longifolia 9(3): 128-131, 128pl, 130d
oblongifolia 9(3): 130
parvifolia 9(3): 130
Australia, Cape York Peninsula 1(3):
122-132
Australasian ferns in Irish gardens
5(3): 129-136
Austrocedrus 12(2): 67
Avocado (*Persea americana*): 1(3): 134
Axonopus paschalis 14(4): 227
Azilia eryngioides 7(3):121-124,122pl,
123d

Bailes, Christopher 2(1): 195-197; 2(2):
242, 273-284; 3(1): 8-13
Baker, William 11(3): 104-109; 13(1): 8-
13
Balfour, I.B. 13(4): 200, 204
Balmisia 7(1): 17
Balsaminaceae 14(1): 23; 15(1): 37
Bamboos 12(1): 8-15
Bambusa 12(1): 9
viridi-glaucescens 12(1): 14
Banks, Sir Joseph 10(2): 85; 14(2): 87-88
Baobabs of Madagascar 13(2): 86-95
Baptisia 14(4): 181
Baptistonia echinata 13(3): 154d, 155,
156d
Barbara's buttons 15(3): 158
Barham, Joan.M. 11(4): 171-174; 13(1):
19-21; 14(4): 203-207
Barker, Christine 11(2): 65-69
Barkeria melanocaulon 5(4): 154-156,
154pl, 155d
Barklya 11(4): 167

Barlia 11(1): 23
Barlow, Gillian 15(2): 104pl, 107d, 154
Barringtonia 9(4): 164, 170-172
edulis sensu Bailey 9(4):174
edulis sensu Guppy 9(4): 174
edulis sensu Seeman 9(4): 166, 167d,
169d, 175-179, 176pl
excelsa Benth. in Hook. 9(4): 174
excelsa sensu Seeman 9(4): 175
guppyana 9(4): 174
magnifica 9(4): 174
novae-hiberniae 9(4): 166, 172pl
procera 9(4): 166, 174pl, 178
schuchartiana 9(4): 174
seaturae 9(4): 176
speciosa 9(4): 174
Barun Khola 1(1): 30-35, 30d, 31m
Bat-winged hedysarum 13(1): 19
Bates, Mary 6(4): 152pl; 7(3): 122pl;
8(1): 18pl, 22pl; 8(2): 64pl, 68pl, 70pl;
11(2): 56pl, 58d; 15(4): 250pl
Batten, Auriol 12(4): 202-206, 202pl
Baudouinia rouxvillei 8(4): 197-202, 200d;
14(4): 234, 235
Bauer Brothers 14(2): 93d, 94pl
Colour Chart 14(2): 92pl
Bauer, Ferdinand: original drawings 3(4):
176-178, 177d; 14(2): 87-100; 15(3):
210, 211; 15(4): 254-272
Bauer, Franz 15(3): 209-211
Bauhinia 14(4): 181
castilloi 11(4): 169
galpinii 11(4): 167
glauca 11(4): 168
guianensis 11(4): 168
jenningsii 11(4): 167-171, 168pl, 170d
monandra 11(4): 167; 14(4): 235
pauletia 11(4): 168
picta 11(4): 167
purpurea 11(4): 168
racemosa 11(4): 168
tomentosa 11(4): 167
vahlii 11(4): 168

variegata 11(4): 167
Beckett, Kenneth 5(2): 82-88; 8(2): 54-88
Begonia 7(2): 57-85
  amphioxus 7(2): 77-81, 78pl, 80d
  bogneri 3(3): 112-116, 112pl, 115d
    'Little Winky' 3(3): 114
  cauliflora 7(2): 68-72, 68pl, 71d
  erythrogyna 7(2): 81-85, 82pl, 83d
  imbricata 7(2): 64-68, 64pl, 67d
  kinabaluensis 7(2): 72-77, 72pl, 75d
  malacosticta 7(2): 61-64, 62pl, 63d
Begonias from Sabah 7(2): 57-85
Bell, Professor E. A. 1(1): 1
Bell, Sandra 7(1): 6-9; 8(1): 5-8; 9(2): 55-
  59; 11(1): 22-28, 32-37; 13(3): 157-164;
  15(3): 169-174, 216
Bellevalia hyacinthoides 11(3): 128-131,
  128pl
  sect. *Strangweia* 11(3): 128
  spicata 11(3)128, 130
Benstead, Mary 7(1): 10pl; 9(2): 54
Berberidaceae 12(1): 15-25
Berg, Hendrieke 15(1): 38pl, 40d, 82
Berne Convention 1(1): 25
Betula 6(2): 65-73; 12(1): 33-35
  collections of 3(1): 40
  chinensis 6(2): 70pl
  utilis 13(2): 101; 14(3): 158, 161
    var. *jacquemontii* 'Grayswood
      Ghost' 6(2): 66pl, 67
  series *Acuminatae* 6(2): 70
  series *Albae* 6(2): 68
  series *Costatae* 6(2): 68
  series *Humiles* 6(2): 70
Biarum 12(3): 141
  auraniticum 12(3): 141
  davisii 12(3): 141
  ditschianum 12(3): 141-144, 142pl, 143d
Bignonia angustifolia 8(1): 22
Billbergia 13(1): 22
  rhodocyanea 1(1): 6
Billiet, Frieda 8(4): 163-167; 13(1): 14-18
Birch (*Betula*): collections 3(1): 40

Birds & Flowers, co-evolution 11(4): 198-
  206
Blandfordia flammea 4(1): 42d
Blepharispermum 13(4): 196
Block, Petra de 15(1): 42-48
Bogner, Josef 12(3): 150-153, 153-168
Boissier, Edmund 11(3): 131-143, 132p.
  137d, 139p
Boissier, Floras & Types: some
  observations 11(3): 143-146
Boissier, Lucile 11(3): 134
Bomarea 14(1): 7, 8
  caldasii 6(4): 176, 176d
  dulcis 6(4): 176
Bomabacaceae 13(2): 86
Bos, Fred 10(4): 190-196
Bossiaea walkeri 11(4): 171-174, 172pl,
  173d
Bossieu de la Martiniéri 11(4): 172
Botanical Drawings, recording colour in
  late 18th century 14(2): 87-100
Botanical Reserves 3(1): 46-48
Botanizing in north-east Greenland
  6(1): 29-37
Bottle Brush Orchid (*Dendrobium
  smilliae*): 1(3): 128
Bottle palm 13(4): 186, 188
Bouvardia, Two cultivated species of
  10(4): 166-170
Bouvardia flava 10(4): 168
  laevis 10(4): 168-169, 168pl, 168d
  mollis 10(4): 168
  nubigena 10(4): 168
  splendens 10(4): 169
  ternifolia 10(4): 169-170, 170pl, 170d
  triphylla 10(4): 169
  subg. *Bouvardia* 10(4): 166
  subg. *Bouvardiastrum* 10(4): 166
  subg. *Bouvardioides* 10(4): 166
Boyce, Peter 5(1): 18-21; 5(2): 72-76;
  5(4): 151-153; 7(1): 14-21; 9(2): 67-75;
  10(3): 121-124; 11(2): 59-65; 12(2): 85-
  89, 122-125, 131-134, 137, 138-140,

141-144, 145-147, 150-153, 153-168, 168-173, 173-176; 14(2): 82-86
Brachystegia 14(4): 192
Bradley, Peter T.J. 8(3): 118-123,155-158
Brady, Aidan 12(4): 199-200
Brandham, Peter 3(1): 57-61; 6(2):58-61; 7(3):124-8
Brauneria purpurea 4(1): 34
  serotina 4(1): 34
Brenierea 11(4): 167
  insignis 14(4): 237, 238d
Brewster, P. 12(3): 168-173
Brighamia citrina 3(2): 69
  var. *napalensis* 3(2): 68-72, 68pl, 71d
  distribution in Hawaii 3(2): 70m
  insignis 3(2): 68
Brodie, Sue 15(1): 2-11, 27-36
Bromelia variegata 9(3): 126
Bromeliaceae 13(1): 22-26
  subfam. *Bromelioideae* 6(1): 24
  subfam. *Pitcairnioideae* 6(1): 24
  subfam. *Tillandsioideae* 6(1): 24
Bromeliads threatened by trade 6(1): 22-29
Bromeliads from South America, Three interesting 8(2): 63-71
Broussonetia 14(4): 228
Brown, Andrew 15(3): 164pl
Brown, Bob 15(2): 115-119
Brownea 14(4): 181
  ariza 12(4): 240
  grandiceps 12(4): 240
Brugmansia sanguinea 6(4): 177, 178d
Brummitt, R.K. 7(4): 161-5
Bruyninckx, Diane 8(4): 164pl; 10(2): 76pl, 80d; 13(1): 14pl, 17d
Bucephalandra 12(3): 131-134, 150-153
  catherineae 12(3): 150-153, 151d
  gigantea 12(3): 150-151
  motleyana 12(3): 131-134, 132pl, 150-152
Buchan, Alexander 14(2): 87-88
Buckley, Samuel Botsford 11(2): 66
Bulbophyllum 13(3): 139

grandiflorum 6(3): 126d
'Bulbous' plants, conservation of 2(2): 241
Bulbs, collecting 3(1): 41
  conservation of 3(1): 23-29
  endangered 3(2): 95-96
Bulley, A.K. 13(2): 108
Bunge, A. von 14(1): 35-55, 47p
Bunya Pine 8(4): 176, 179
Bupthalmum 13(4): 195
Burren, The, County Clare, Ireland 2(3): 312-320, 312m
Burton, Sir Frederick 13(1): 37
Bustard, Louise 7(2): 85-90
Butea 14(4): 176, 182
Butini, Lucile 11(3): 134
Butonica edulis 9(4): 175
  procera 9(4): 174
  samoensis 9(4): 176
Butterfly plant 13(1): 19
Bytebier, B. 15(1): 49-62

Caatinga 13(2): 70-71, 75-76
  plants 2(1): 216
  typical scenery, during rainy season 2(1): 217d
Cactaceae 11(2): 70; 13(2): 70; 14(3): 125; 15(2): 96
  tribe *Cereeae* 12(2): 70, 72, 75
  tribe *Rhipsalideae* 14(3): 125-126
  subtribe *Borzicactinae* 15(2): 103
Cacti, conservation of 1(4): 181-187
Cactus, Hedgehog see *Echinocereus*
Caesalpinia 14(4): 176, 181, 187, 242, 243-244
  ferrea 2(1): 224, 224d
  major 14(4): 227
Caesalpiniaceae 14(4): 179
Cafferty, Steve 13(4): 210-214; 15(4): 236-241
Cajanus 14(4): 189
Calanthe triplicata 6(3): 131d
  vestita 13(3): 159
California Native Plant Protection Act

1(1): 27
Calliandra 14(4): 176, 181
  depauperata 2(1): 223-224
Callipteris accedens 8(3): 131
  attenuata 8(3): 131
  prolifera 8(3): 131
  spinulosa 8(3): 131
Callirhoe involucrata 11(3): 150
  var. *lineariloba* 11(3): 150
  leiocarpa 11(3): 150
  scabriuscula 11(3): 146-151
Calocedrus 12(2): 66-67
Calophyllum inophyllum 13(4): 222
Calotropis gigantea 9(4): 185d
Calypso bulbosa 3(4): 147-148
  f. *occidentalis* 3(4): 150
  var. *japonica* 3(4): 151
  var. *occidentalis* 3(4): 147, 148-150
    148pl, 149d
  var. *speciosa* 3(4): Cover pl
  japonica 3(4): 150, 151
  occidentalis 3(4): 150
  speciosa 3(4): 150, 151
camera clara 15(4): 265
camera lucida 15(4): 265-266, 267-271,
  269fig.
Camera obscura 15(4): 265-266
Campanulaceae 13(4): 200-203
  tribe *Lobelioideae* 13(4): 210
Camphor Tree (*Cinnamomum camphora*):
  1(3): 133
Campos rupestres 13(1): 33; 13(2): 70-71
Camptosema scarlatinum var. *pubescens*
  9(1): 12-14, 12pl
Campylotropis 14(4): 203-207
  macrocarpa 14(4): 203
  polyantha 14(4): 203-207, 204pl, 205d
Canarium indicum 6(3): 131, 132
Candolle, A.-P. de 11(3): 133-134
Cape York Peninsula 1(3): 122-132
Capparis 13(2): 86
Caprifoliaceae 14(3): 120
Caprifolium pilosa 10(3): 112

Caragana 14(4): 180, 182
Caraguata multiflora 8(2): 65
Cardiocrinum cordatum var. *glehnii* 2(1):
  232
Carduus subcoriaceus 6(4): 166
Carex 11(1): 6
Carmichaelia 12(4): 182
Carpinus turczaninowii var. *ovalifolia*
  3(3): 137
Casparea jenningsii 11(4): 169
Cassia 14(4): 176, 181
  candolleana 5(2): 79
  closiana 5(2): 79
  corymbosa Lam. 5(2): 77
  corymbosa sensu Sims 5(2): 77
  floribunda sensu Lowe 5(2): 80
  floribunda sensu de Wit 5(2): 79
  laevigata 5(2): 79
  obtusa 5(2): 79
  septentrionalis 5(2): 79
Castanospermum 14(4): 189, 243
Catasetum roseum 7(1): 9
Catherine, Eleanor 12(2): 65d; 12(3):
  151d, 152; 14(3): 152d
Cattleya citrina 2(1): 195
  doweana 11(1): 35
  forbesii 13(3): 159
Caujapé-Castells, Juli 15(1): 67-77
Cedrela glabra 15(3): 189
  longifolia 15(3): 189
    var. *kumaona* 15(3): 189
  serrata 15(3): 189
  serrulata 15(3): 189
  sinensis 15(3): 189
    var. *hupehana* 15(3): 190
    var. *lanceolata* 15(3): 190
    var. *schensiana* 15(3): 190
Centaurea 13(1): 3-4
  abyssinica 13(1): 4
  americana 13(1): 3
  chilensis 13(1): 3
  floccosa 13(1): 4
  grandiflora 13(1): 5

9

mexicana 13(1): 3
rothrockii 13(1): 5
sect. *Plectocephalus* 13(1): 4
Centropogon ?*andropogon* 9(1): 5
  bonplandianus 9(1): 5
    var. *intermedius* 9(1): 5
  cornutus 9(1): 3-7, 4pl
    var. *angustifolius* 9(1): 5
    var. *intermedius* 9(1): 5
  edulis 9(1): 5
  fastuosus 9(1): 5
  intermedius 9(1): 5
  laevigatus 9(1): 5
  oblongus 9(1): 5
  puerilis 9(1): 5
  surinamensis 9(1): 5
    var. *angustifolius* 9(1): 5
    var. *vestitus* 9(1): 5
Centrosis abortiva 10(4): 164
Centrostemma coriaceum 7(1): 6
  cyrtoceras 7(1): 6
  lindleyanum 7(1): 6
  multiflorum 7(1): 6
Cephalanthera 1(2): 88-90
  royleana 10(1): 20
Cephalocereus penicillatus 13(2): 77
Ceratonia oreothauma 2(4): 380-382
Cercis 14(4): 181
Cereus acanthurus 15(2): 98
  acifer 1(4): 157, 159
  chloranthus 1(4): 169
  engelmannii 2(2): 246
    var. *chrysocentrus* 2(2): 247
    var. *variegatus* 2(2): 248
  eriotrichus 15(2): 98
  fendleri 2(2): 251
  glaber 2(4): 341
  munzii 2(2): 247
  pectinatus 1(4): 174
  penicillatus 13(2): 77
  phoeniceus 1(4): 160
  poselgeri 2(2): 265
  pulchellus 2(2): 272

scheerii 1(4): 150-151, 154
stramineus 1(4): 163
tonduzii 2(4): 341
tuberosus 2(2): 265
viridiflorus 1(4): 169
Ceropegia arabica
  var. *abbreviata* 8(1): 31, 32d
  var. *arabica* 8(1): 31, 32d
  aristolochioides subsp. *deflersiana*
    8(1): 28, 29, 29d, 36; 8(4): 205
  botrys 8(1): 29d 36
  bulbosa 8(1): 33, 34, 34d
  superba 8(1): 31, 32d
  superba x *arabica* var. *abbreviata* 8(1):
    32
  tihamana 8(1): 29d
  variegata var. *adelaidae* 8(1): 26, 27,
    27d, 36
    var. *variegata* 8(1): 26, 27d
  vignaldiana 8(1): 34, 34d, 35
Ceropegias in Saudi Arabia 8(1): 24-37
Cerrado de altitude 13(2): 70
Chamaedorea 134): 187
  metallica 4(4): 160-4,
    160pl, 163d
Chamaenerion latifolium 6(1): 35d
Chamaepericlymenum suecicum 6(2): 80
Chamaerops martiana 11(2): 77
Chameleo fischeri 3(2): 85, 85d
Changnienia amoena 10(1): 52-54,53d
Chase, Mark W. 13(3): 130-133
Cheek, Martin 5(2): 60-65; 10(3): 138-
    144; 11(2): 55-69, 88-94; 13(4): 210-
    214; 15(1): 27-36; 15(4): 236-241
Cheetham, G. E. 1(4): 147
Cheiroglossa malgassica 13(4): 233
Chelidonium fulvum 6(4): 151
  glaucium 6(4): 151
  littorale 6(4): 151
Chesneya 14(4): 182
Chickpea 14(4): 189
Chile, The Alerce Andino National
  Park 4(3): 127-140

Chilean blue crocus 15(2): 126-127
Chinese mahogany 15(3): 188
Chionodoxa luciliae 11(3): 134
Chitwan National Park, Nepal 9(4):
   183-193
Choiseul, Orchid Hunting in 6(3):
   124-134
Choisya arizonica 15(3): 176, 178
  'Aztec Pearl' 15(3): 175, 177
  dumosa var. *arizonica* 15(3): 175-179,
    176pl, 178d
    var. *dumosa* 15(3): 176
    var. *mollis* 15(3): 176
  ternata 15(3): 175, 177
    'Sundance' 15(3): 175
  ternata x *dumosa* var. *arizonica* 15(3):
    175
Chordospartium 14(4): 182
Christia 13(1): 19-21
  lunata 13(1): 20
  vespertilionis 13(1): 19
    var. *grandifolia* 13(1): 20
    var. *vespertilionis* 13(1): 19-21, 20pl,
    21d
Christmas cactus 14(3): 125
Chrysanthemum arcticum
  subsp. *maekawanum* 5(2): 59
  miyamojimense 5(2): 59; 6(1): 48
  yezoense 5(2): 59
    var. *lobulifolium* 5(2): 59
    var. *stipulatum* 5(2): 59
    var. *typicum* 5(2): 59
Chuquiragua jussieui 6(4): 175d
Cicer 14(4): 189
Cinnamomum camphora 1(3):133;
   4(2): 80d
  glanduliferum 1(3): 134, 135
  zeylanicum 6(4): 166
Cipocereus 13(2): 71-73
  bradei 13(2): 73
  crassisepalus 13(2): 71
  laniflorus 13(2): 71
  minesis 13(2): 73

  pusilliflorus 13(2): 73
Cirsium heterolepis 6(4): 166
  maximum 6(4): 166
  pinnatisectum 6(4): 166
  subcoriaceum 6(4): 163-166, 164pl, 165d
Cissus discolor 13(4): 200
CITES (Convention on International Trade
  in Endangered Species) 1(4): 181-187;
CITES, and Nepenthes 12(2): 113-118
Cladrastis 14(4): 182
Claoxylon linostachys 13(4): 233
Clarke, G.P. 15(1): 62-67
Cleistocactus acanthurus 15(2): 96-103,
  96pl, 101d
  var. *faustianus* 15(2): 100
  var. *pullatus* 15(2): 100
  sextonianus 15(2): 96pl, 98, 101d
Clematis confusa 8(4): 159-162, 160pl,
  162d
  connata 8(4): 161
  glycinoides 5(4): 175d
  marmoraria 4(3): 116-119, 116pl, 118d
Clements, Mark 1(2): 50, 84-91; 1(3):
  139-42; 3(1): 8-13
Clerodendron viscosum 9(4): 187d
Clethra barbinervis 10(4): 188d
Clianthus 14(4): 182
  carneus 15(2): 127
  formosus 1(1): Cover pl
Clitoria 14(4): 182
  fulgens 9(1): 14
Clowesia rosea 7(1): 6-9, 6pl, 8d
Clusiaceae (Guttiferae) 15(2): 92
Cnicus heterolepis 6(4): 166
  pinnatisectum 6(4): 166
  subcoriaceus 6(4): 166
Coccocypselum 10(2): 76
  guianense 10(2): 76-80, 76pl, 80d
  hirsutum 10(2): 77
Coelogyne mooreana 12(4): 239
  pandurata 11(1): 36
Coevolution of birds and flowers 11(4):
  198-206

11

Cola 11(2): 89
Colenso, William, New Zealand
  Botanist 7(4): 186-200
Colensoa physaloides 7(4): 197d
Coleocephalocereus 13(2): 71-76
Collecting: danger to conservation 3(3): 97
Collections: aiding endangered species
  3(2): 49
(Living) Collections at Kew 3(1): 39-48
Collaea scarlatina var. *pubescens* 9(1): 14
Collenette, Sheila 5(1): 36-8; 8(1): 24-37;
  14(1): 31-34
Colophospermum mopane 14(4): 192
Colour Charts, Bauer Brothers 14(2): 87-
  100
Colour Charts, a modern artist's view
  14(2): 100-101
Colutea 14(4): 180, 182
Colvillea 14(4): 235
  racemosa 14(4): 235, 236d
Commercial exploitation of plants 2(3):
  289-290
Commidendrum burchellii 2(4): 375
  robustum 2(4): 374
  rotundifolium 2(4): 374-375
  rugosum 2(4): 373d
Compositae 12(1): 5-8; 13(1): 3-7; 13(2):
  56; 13(4): 191-199; 15(3): 158
  tribe *Gnaphalieae* 13: (2): 57; 13(4): 196
  tribe *Heliantheae* 13(4): 195
  tribe *Inuleae* 13(2): 57; 13(4): 195-196
  tribe *Inuloideae* 13(4): 195
  tribe *Plucheeae* 13(4): 196
  tribe *Relhaniinae* 13(2): 57
  tribe *Vernonieae* 13(4): 191
Compton, James 11(2): 52-55
Condy, Gillian 11(1): 10pl, 12d; 14(1):
  16pl
Confusing Cassias 5(2): 76-82
Congo Republic, lists of orchids 13(3):
  169-176
  map of 13(3): 165, 166
Conservation 1(3): 97-99; 1(4): 145-146;
  3(1): 3
  of cacti 1(4): 181-187
  of *Cypripedium calceolus* 1(3): 141-142
  of endangered bulbs 1(1): 23-29
  of habitats 2(3): 289
  of *Rhizanthella* 1(2): 90-91
  of Saintpaulia 15(1): 49-62
  of the toromiro tree 14(4): 226-231
Conservation in Nepal 1, the
  Langtang National Park 5(1): 24-32
Conservation in Nepal 2, the Royal
  Chitwan National Park 9(4): 183-193
Conservation in Nepal 3, the Rara National
  Park 13(2): 96-104
Conservation in Nepal 4, Mt Everest
  National Park 14(3): 153-166
Convention on International Trade in
  Endangered Species (CITES): 1(4):
  181-187
Conyza populifolia 13(4): 192
Cook, Captain James 12(4): 186; 14(2):
  87-88; 14(4): 222, 227
Cook, Linda 15(2): 86-91
Cooke, David N. 4(4): 160-164; 14(4):
  198-202
Coombes, Allen J. 7(3): 133-138
Coombs, Jill 6(2): 66pl, 70pl
Cordyla 14(4): 234, 235
Cornus 6(2): 74-83
  'Ascona' 8(2): 74
  'Eddie's White Wonder' 8(2): 74
  capitata 7(4): 179-181
  florida 8(2): 75d
  japonica 7(4): 176
    'Norman Hadden' 7(4): 179-181
    'Porlock' 7(4): 179-181, 180pl
  nuttallii 8(2): 71-77, 75d
    in cultivation 8(2): 71-78
    'Colrigo Giant' 8(2): 72
    'Colrigo Wonder' 8(2): 72
    'Eddiei' 8(2): 73
    'Gold Spot' 8(2): 73
    'Monarch' 8(2): 73

'North Star' 8(2): 73
'Pilgrim' 8(2): 73
'Portlemouth' 8(2): 73
nuttallii x *florida* 8(2): 76-77
'Ormonde' 8(2): 74-76, 75d
Coresantha alata 3(2): 65
Coronilla 14(4): 182
 glauca 14(4): 183
Correa sp. 15(4): 263d
Corydalis 15(1): 14
 balsamiflora 15(1): 16
 elata 15(1): 12
 flexuosa 15(1): 12-19, 12pl, 16d
  'Blue Dragon' 15(1): 14
  'Blue Panda' 15(1): 13
  'China Blue' 15(1): 14
  'Balang Mist' 15(1): 15
  'Nightshade' 15(1): 14
  'Père David' 15(1): 14
  'Purple Leaf' 15(1): 14
  'Purple Panda' 15(1): 14
  subsp. *flexuosa* 15(1): 12
   var. *flexuosa* 15(1): 12
   var. *microflora* 15(1): 12
   var. *mucronipetala* 15(1): 12
   var. *omeiana* 15(1): 12
   var. *pinnatibracteata* 15(1): 12
 popovii 5(2): 65-9, 66pl, 68d
 sect. *Leonticoides* 5(2): 65
Corynocarpus laevigatus 3(3): 125-126,
 125d
Coryphantha minima 1(4): 186
 scheeri 11(2): 80
Corsia, the saprophytic genus in the
 Solomon Islands 2(3): 320-323
 haianjensis 2(3): 312d, 322, 323;
 pyramidata 2(3): 321d, 322, 323;
Costia scorpioides 3(2): 65
Cotoneaster bradyi 12(4): 198-201, 198pl,
 210d
 series *Dielsiani* 12(4): 199
Cowley, Jill 1(1): 2, 11-14; 2(2): 285-288;
 3(3): 116-118; 4(4): 199-201; 5(1): 11-

15; 6(2): 61-4, 6(3): 95-101; 11(1): 13-
18; 11(3): 104-109; 12(2): 73-77; 13(1):
8-13; 13(2): 108-111; 14(1): 2-6; 14(2):
77-81; 15(4): 220-225, 226-230
Crassocephalum auriformis 7(3): 112
 ruwenzoriensis 7(3): 112
 scandens 7(3): 112
Crawfordia japonica 2(1): 227
Cress-Rocket (*Vella pseudocytisus*): 3(3):
 140, 141d
Cretan *Daphne*, A close look at a 8(2):
 37-45
Cribb, Phillip 1(2): 50, 84-91; 1(3):
 122-132; 2(3): 291-308, 320-323; 2(4):
 351-354; 3(1): 4-13; 3(2): 55-57, 66-67;
 3(3): 102-103; 3(4): 159-166; 4(4): 157-
 160; 5(1): 22-24; 5(4): 161-167; 7(3):
 117-120; 10(1): 5-8, 13-18, 36, 52-54;
 12(2): 62-66; 13(3): 152-157; 14(3):
 148-153
Crinodendron hookerianum 4(3): 130d
Crinum moorei 12(4): 241
Crocker, Miss Emmeline 10(4): 177
Crocus 3(4): 187, 189; 15(2): 120-121
 abantensis 1(2): 69
 aureus 1(2): 72
 baytopiorum 1(2): 69-70, 70d
 biflorus 1(2): 69; 15(2): 121
 candidus 3(4): 188pl
 dalmaticus 7(4): 184
 gargaricus 1(2): 70-73
  subsp. *gargaricus* 1(2): 72, 72d
  subsp. *herbertii* 1(2): 72, 72d
 goulimyi 1(2): 73; 1(4): 192
  f. *albus* 1(2): 73; 1(4): 192
 hadriaticus 2(3): 309-311, 310d
  f. *hadriaticus* 2(3): 311
  f. *lilacinus* 2(3): 311
  f. *parnassicus* 2(3): 310-311
  var. *saundersianus* 2(3): 310
 kerndorffiorum 15(2): 120-124
 kotschyanus 1(4): 192
 leichtlinii 15(2): 121-122

pelistericus 1(2): 74
rujanensis 7(4): 182-186
scardicus 1(2): 73-74
speciosus 1(4): 192
thirkeanus 1(2): 72, 73
zonatus 1(4): 192
Crocuses in cultivation, Rare and little
  known 1(2): 68-74
Crocus from Yugoslavia, A new species of
  7(4): 182-186
Croix, Isobyl la 14(3): 166-169; 15(1): 20-
  26; 15(3): 164-174; 15(4): 249-254
Crotalaria 14(4): 177, 181, 208-212
  anisophylla 14(4): 209
  ekmanii 14(4): 208
  heterophylla 14(4): 208
  quinquefolia 14(4): 208-212, 208pl, 211d
  recta 14(4): 210
  retusa 14(4): 210
  spectabilis 14(4): 210
  urbaniana 14(4): 209
  sect. *Crotalaria* 14(4): 209
Crown Imperial 13(1): 28-29
Cryptanthemis slateri 1(2): 84-91
Cryptostephanus herrei 11(1): 10, 12
Ctenitis canacae 13(4): 233
  desvauxii 13(4): 233
Cuffe, Lady Charlotte Wheeler 12(4): 243
Culcasia scandens 12(3): 138
  seretii 12(3): 138-140, 138pl, 139d
Culham, Alastair 15(2): 124-132
Cullen, James 1(3): 99, 119-122
Cultivation, of Aroids at Kew 12(4): 168-
  173
Cultivation of woody legumes under glass
  at Kew 14(4): 242-244
Cuminia brevidens 3(4): 156
  eriantha 3(4): 151-6, 152pl, 155d
    var. *eriantha* 3(4): 154, 155d
    var. *fernandezia* 3(4): 153, 154, 155,
      155d
  fernandezia 3(4): 152, 156
Cunningham, Allan 5(4): 167-180

Cupressaceae 11(4): 156
Cupressus dupreziana 3(2): 88-90, 89d
  cashmeriana 11(4): 156-166, 156pl,
    160d,164d
  corneyana 11(4): 157-159
  dupreziana 11(4): 156
  funebris 11(4): 156, 163
    var. *glauca* 11(4): 161-163
  himalaica 11(4): 159, 161, 163
    var. *darjeelingensis* 11(4): 163
  pendula 11(4): 157, 163
  sempervirens 11(4): 156
  torulosa 11(4): 158-159, 162
    var. *cashmeriana* 11(4): 163
  torulosis 11(4): 157-159, 163
Curtis, William 2(4): 368; 4(1): 7-14
  at Chelsea Physic Garden 4(1): 51-56;
    13(1): 1p
Curtisia dentata 2(4): 368
Curtis's Botanical Magazine, bicentenary
  4(1): 3
  and Ninian Niven 15(4): 274-282
  facsimile of Vol.1, 4(1): 15-31
  Hibernian contribution 4(1): 39-51
Cyathostegia 14(4): 187, 192
Cyclamen colchicum 3(3): 107-108; 15(3):
    180-185, 180pl, 184d
  coum 3(3): 95-96; 4(1): facsimile t.4, 35,
    36-38
  europaeum 3(3): 105-106, 109
    subsp. *ponticum* 15(3): 183
    var. *colchicum* 15(3): 183
    var. *ponticum* 15(3): 183
  fatrense 3(3): 107, 109
  hederifolium 1(1): 25, 26; 3(3): 105
  mirabile 1(1): 26, 26d; 3(2): 95
  ponticum 15(3): 184
  purpurascens 3(3): Cover pl, 104-109,
    106pl
    subsp. *ponticum* 15(3): 184
Cylindrocline 13(4): 195-199
  commersonii 13(4): 195-199, 196d,
    196pl, 198d

14

lorencei 13(4): 195
Cymbidium aloifolium 9(4): 190d
  devonianum 1(2): 78
  finlaysonianum 13(3): 161
  insigne 1(2): 78, 79d
  lancifolium 1(2): 80-81
  rectum 11(1): 35
Cymbiglossum cervantesii 5(2): 55
Cynometra 14(4): 234, 235
Cynorchis 13(3); 124
  macloughlinii 6(1): 17
Cynorkis 13(3): 124-129
  fastigiata 13(3): 124
  gibbosa 13(3): 124-129, 124pl, 128d
  gigas 13(3): 124
  grandiflora 13(3): 124
  lowiana 13(3): 124
  purpurascens 13(3): 124
  uncinata 13(3): 124
  uniflora 13(3): 124
  x *kewensis* 13(3): 124
Cynosorchis 13(3): 124, 126
Cypella caerulea 9(1): 8
Cyperaceae 11(1): 6
Cyperus aequalis 11(1): 7
  haspan 11(1): 7
  haspan viviparus 11(1): 7
  involucratus 11(1): 6
  isocladus 11(1): 7
  papyroides 11(1): 7
  papyrus 11(1): 6
  prolifer 11(1): 6-9, 6pl, 8d
    var. *isocladus* 11(1): 6
  rotundus 11(1): 6
Cypress, Tassili (*Cupressus dupreziana*):
  3(2): 88-90, 89d
Cypripedium calceolus 1(3): 139-142,
  140d; 10(1): 42; 11(1): 32; 14(3): 148
  chinense 14(3): 149, 151
  exul 10(1): 7
  fairrieanum 2(4): 354
  formosanum 4(2): 64
  henryi 14(3): 148-153, 148pl, 152d

insigne 10(1): 7
japonicum 3(1): 44d
  var. *formosanum* 4(2): 62-64, 62pl, 63d
japonicum sensu Hayata 4(2): 64
sanderianum 1(1): 6
segawai 14(3): 148
Cyrtanthus breviflorus, the flowering of
  after a fire 6(1): 38-39
  carneus 11(1): 11
  elatus 11(1): 9
  herrei 11(1): 9-13, 10pl, 12d
  mackenii 11(1): 9
  purpureus 11(1): 10
Cyrtoceras coriaceum 7(1): 6
  floribundum 7(1): 6
  lindleyanum 7(1): 6
  multiflorum 7(1): 6
  reflexum 7(1): 5
Cytherea bulbosa var. *occidentalis*
  3(4): 150
Cytisus x *dallimorei* 14(4): 183

Dactylorhiza fuchsii 11(1):24
Dadd, Richard 4(2): 91-96
Daguerre, L.J.M. 15(4): 271
Dalbergia 14(4): 176, 182, 192, 234, 235
Dalea 14(4): 181, 182
Dallimore, W. 10(2): 89, 90
Dalström, Stig 6(3): 135pl
Dawn redwood 15(1): 77
Danthonia paschalis 14(4): 227
Daphne cannabina var. *kiusiana* 6(3): 115
  gnidioides 8(1): 37-42, 42d; 11(3): 117-
    119, 121map, 122
  jasminea 11(3): 119-128, 124pl, 126d
  kiusiana 6(3): 112-115, 114pl, 114d
    f. *fasciata* 6(3): 114
    var. *atrocaulis* 6(3): 114
  mezereum 11(3): 124
  microphylla 11(3): 124
    var. *angustifolia* 11(3): 124
  odora var. *atrocaulis* 6(3): 114
    var. *kiusiana* 6(3): 115

15

oleoides 11(3): 117, 119, 121-122, 124-125

oleoides var. *jasminea* 11(3): 124

oleoides sensu d'Urv. 8(1): 43

sericea 11(3): 117

Darwin, Charles 13(3): 161

Dean, Pauline 11(1): 4pl, 5d; 14(1): 24pl, 28d; 15(2): 120pl

de Block, P. 15(1): 42-48

Dees, Mary 10(3):144-148

Delonix 14(4): 176, 181, 187, 232

  boiviniana 14(4): 233d

  regia 14(4): 183, 233d, 235

Dendranthema miyatojimense 5(2): 59; 6(1): 48

  yezoense 5(2): 55-60, 56pl, 57d

Dendrobium 13(3): 139-141

  agathodaemonis 2(3): 297

  alexandrae 7(3): 117-120, 118pl, 119d

  amplum 13(3): 139-140

  anosmum 13(3): 159

  asperifolium 2(3): 297

  atromarginatum 2(3): 297

  bellum 3(2): 57

  bifalce 1(3): 127

  bigibbum 2(3): 293

  ?*brachycarpa* 11(2): 55

  calcaratum 2(1): 211

  canaliculatum 1(3): 128, 130

    var. *nigrescens* 1(3): 128

  carronii 1(3): 129d, 130

  crumenatum 13(3): 161

  cuthbertsonii 2(3): 294, 295, 296d, 296pl, 298

  discolor 1(3): 126-127, 128

  elongatum var. *orientale* 13(3): 143

  eximium 3(2): 55-7, 56pl, 56d

  fargesii 13(3): 139-140

  gouldii 6(3): 127d; 10(1): 13-18, 14pl, 17d

    var. *acutum* 10(1): 16

    'Guadalcanal Gold' 10(1): 14

  hellwigianum 2(3): 294, 298-299, 298pl, 299d

  hookerianum 2(3): Cover pl

  jenkinsii 13(3): 159

  laevifolium 2(3): 296

  lichenicola 2(3): 297

  lineale 2(3): 293; 10(1): 14

  macrophyllum 2(3): 293, 304

  madonnae 2(3): 304, 305

  microblepharum 2(3): 300, 301

  pulchellum 11(4): 191, 192d

  rarum 2(1): 211-13, 212pl, 213d

  rhodostictum 2(3): 304-305, 304pl, 305d

  schulleri 11(1): 34

  smilliae 1(3): 128, 130, 132

  sociale 11(1): 33

  sophronites 2(3): 295, 297

  spectabile 2(3): 304

  spectatissimum 2(2): 280

  tapiniense 2(3): 306-307, 306pl, 307d

  trachyphyllum 2(3): 297

  undulatum var. *woodfordianum* 10(1): 16

  unicum 1(2): 75; 11(1): 33

  vexillarius 2(3): 300-301, 300pl, 301d

  violaceum 2(3): 294, 302-303, 303pl, 303d

  wollastonii 3(2): 57

  sect. *Cuthbertsonia* 2(3): 294

  sect. *Gastridium* 6(3): 129

  sect. *Latouria* 2(3): 294

  sect. *Oxyglossum* 2(3): 294

  sect. *Pedilonum* 2(1): 211

Dendrocalamus 12(1): 9

Desbordesia glaucescens 13(3): 168

Desmodium 14(4): 182

Desmond, Ray 4(1): 7-14; 9(1): 21-27

Desmotrichum 13(3): 139

Deutzia purpurascens 12(4): 243

  'Alpine Magician' 10(4): 171-178, 174pl; 12(4): 243-244

Dialium 14(4): 234, 235

Diascia personata 3(3): 99-100

16

rigescens 3(3): 99-102, 100pl, 101d
  var. *angustifolia* 3(3): 101
  var. *bractescens* 3(3): 102
Dichromena ciliata 10(3): 117
  ciliata var. *cinnamomea* 10(3): 120
  ciliata var. *vahliana* 10(3): 120
  nervosa 10(3): 120
  obtusifolia 10(3): 120
  persooniana 10(3): 120
  pilosa 10(3): 120
  pura 10(3): 120
  tenuifolia 10(3): 120
Dichrostachys 14(4): 234
Dicksonia antarctica 5(3): 130d
  arborescens 2(4): 378
Dictamnus creticus 11(3): 113
Dictyosperma 13(4): 186-187
Diestelkamp, Edward 12(4): 209-220
Dillenia alata 14(2): 88, 89d, 90pl
Dillon, Helen 8(1): 12-15
Dimorphanthera 1(1): 10
  declinata 1(1): 11
  kempteriana 1(1): 9-11, 9d, 10pl
  tridens 1(1): 11
Dionaea muscipula 3(1): 42d, 42
Dionysia
  sect. *Anacamptophyllum* 5(1): 9
    subsect. *Tapetodes* 5(1): 9
  aretioides 2(1): 205
  denticulata 5(1): 7-11, 8pl, 10d
  involucrata 2(1): 205-207, 204pl, 206d
Diospyros diversifolia 13(4): 217
Dipladenia illustris 9(3): 133
  var. *glabra* 9(3): 133
Diplazium accedens 8(3): 129-131
  proliferum 8(3): 128-133, 128pl, 132d
  repandum 8(3): 131
  spinulosum 8(3): 131
    var. *novoquineensis* 8(3): 131
  swartzii 8(3): 131
Diplocaulobium 13(3): 141
Dipterocarpaceae 12(2): 105
Disa cardinalis 5(1): 15-18, 16pl, 17d

'Kirstenbosch Pride' 5(1): 16pl
  graminifolia 13(3): 161
  tripetaloides 13(3): 161
  uniflora 13(3): 161
Discovery and introduction of
  Allium giganteum 4(2): 91-96
Dissotis 14(3): 168-169
  melleri 14(3): 169
  princeps 14(3): 168d, 169
Distephanus 13(4): 191-194
  populifolius 13(4): 191-194, 192pl, 193d
Diuris spp. 13(3): 162
Dodecadenia grandiflora 1(3): 136
Dodecatheon hendersonii 14(3): 137-141,
    138pl, 140d
  h. subsp. *cruciatum* 14(3): 137, 139
  h. subsp. *hendersonii* 14(3): 137
  h. subsp. *parviflorum* 14(3): 137
  integrifolium var. *latifolium* 14(3): 140
  latifolium 14(3): 140
  meadia var. *hendersonii* 14(3): 140
Dog's Tooth Violet 4(1): 38
Dogwood 6(2): 74-83
Dogwood, Pacific 8(2): 71
  Western Flowering 8(2): 71
Dombeya erythroxylon sensu H.Andrews
    15(1): 29, 33
Donald, Duncan 4(1): 51-56
Doneraile, Lady 12(4): 220-236, 221p
Doricera 13(4): 204, 206
  trilocularis 13(4): 204
Dorpat Botanic Garden 14(1): 33-55
Dorr, Laurence J. 8(4): 197-202; 11(3):
    146-151
Dorstenia 4(2): 97
  foetida 4(2): 97
Dortmanna blanda 10(2): 73
  bridgesii 10(2): 73
  lucaeana 10(2): 73
Dortmannia tupa 5(4): 161
Dracophyllum secundum 5(4): 173d
Dracula 13(3): 120-123
  chimaera 13(3): 120

ubangina 13(3): 120
vampira 13(3) 120-123, 120pl, 122d
  'Morgan le Fay' 13(3): 121
  'Wyld Court' 13(3): 121
Dransfield, John 4(4): 160-164; 7(2):
  90-95; 10(2): 81-84; 13(4): 186-190
Drapetes ericoides 2(2): 283
Dressler, Stefan 14(3): 130
Dring, Dr D.M. 10(3): 133
Drosera anglica x *spatulata* 10(3): 141
  tracyi x *D. filiformis* 10(3): 142
  x *californica* 10(3): 142-143
  x *nagamotoi* 10(3): 141-142
  'anpill' 10(3): 143
  'anterim' 10(3): 143
  'California Sunset' 10(3): 143
  'linglica' 10(3): 143
  'linpill' 10(3): 143
  'linthulata' 10(3): 143
Dryas integrifolia 6(1): 32d
Drynaria lomarioides 12(2): 91, 93
Dryopteris levingei 12(4): 197
  quilinoides 13(4): 233
Dunlevy, Mairead 12(4): 220-236
du Noyer, G.V. 15(4): 281
Duperrea 15(1): 43
  insignis 15(1): 45
  pavettifolia 15(1): 42-48, 42pl, 47d
  var. *kontumensis* 15(1): 46
  var. *scabra* 15(1): 46
  forma *mollissima* 15(1): 46
Du Puy, Blaise 13(2): 86-95, 88d
Du Puy, David J. 1(2): 50, 75-84; 13(3):
  124-129, 163-164; 14(4): 231-241

Easter Island 14(4): 221, 226-231
Eastwood, A. 15(1): 49-62
Eccremocactus bradei 2(4): 341
Echeveria scheerei 11(2): 80
Echinocactus 1(4): 185
  pectinatus 1(4): 179
  pulchellus 2(2): 272
Echinocereus 1(4): 147-149

some choice cultivated 2(2): 243-273
abbeae 2(2): 253
acifer 1(4): 157, 159
adustus 2(2): 265-268
  var. *adustus* 2(2): 268
  var. *schwarzii* 2(2): 266pl, 267d, 268
albiflorus 2(2): 252
amoenus 2(2): 272
bonkerae 2(2): 253
boyce-thomsonii 2(2): 253
chloranthus 1(4): 165-170
  var. *chloranthus* 1(4): 170
  var. *cylindricus* 1(4): 169, 186
  var. *russanthus* 1(4): 166pl, 167, 168d,
    169
chisoensis 2(2): 258-261
  var. *chisoensis* 2(2): 260, 261
  var. *fobeanus* 2(2): 259, 260d, 260pl
engelmannii 2(2): 243-248, 244pl
  var. *acicularis* 2(2): 246
  var. *armatus* 2(2): 247
  var. *chrysocentrus* 2(2): 247
  var. *engelmannii* 2(2): 244pl, 245d, 246
  var. *howei* 2(2): 247
  var. *munzii* 2(2): 247
  var. *nicholii* 2(2): 247
  var. *purpureus* 2(2): 248
  var. *variegatus* 2(2): 248
conglomeratus 1(4): 163
cucumis 1(4): 151
dasyacanthus 1(4): 176-177, 179
enneacanthus var. *stramineus* 1(4): 163
fasciculatus 2(2): 253
  var. *bonkerae* 2(2): 253
  var. *boyce-thompsonii* 2(2): 253
fendleri 2(2): 248-254
  var. *bonkerae* 2(2): 253
  var. *boyce-thompsonii* 2(2): 249, 253
  var. *fasciculatus* 2(2): 244pl, 250, 252d
  var. *fendleri* 2(2): 254
  var. *kuenzleri* 2(2): 250, 254
  var. *ledingii* 2(2): 253
  var. *rectispinus* 2(2): 250, 254

var. *robustus* 2(2): 253
ferreirianus 2(2): 252d
fobeanus 2(2): 261
gentryi 1(4): 154
hempelii 2(2): 250, 254
kuenzleri 2(2): 254
ledingii 2(2): 253
leucanthus 11(2): 71
luteus 11(2): 73
madrensis 2(2): 268
maritimus 2(2): 244, 245d
'*matthesianus*' 1(4): 155
nivosus 1(4): 164
ochoterenae 11(2): 74
pacificus 1(4): 160
pectinatus 1(4): 171, 175-180
  var. *ctenoides* 1(4): 180
  var. *dasyacanthus* 1(4): 170pl, 173d,
    176, 179
  var. *neomexicanus* 1(4): 179
  var. *pectinatus* 1(4): 176, 179
  var. *rigidissimus* 1(4): 174
  var. *rubispinus* 1(4): 175
  var. *wenigeri* 1(4): 179
pentalophus 11(2): 70-71
  var. *leonensis* 11(2): 71
  var. *pentalophus* 11(2): 71
polyacanthus 1(4): 155-160
  var. *densus* 1(4): 153d, 156pl,158, 159
  var. *pacificus* 1(4): 157, 160
  var. *polyacanthus* 1(4): 156
poselgeri 2(2): 262-265, 262pl, 264d;
  11(2): 71
pulchellus 2(2): 268-273
  var. *amoenus* 2(2): 269, 271d, 272
  var. *pulchellus* 2(2): 269, 270pl,
    271d, 273
  var. *weinbergii* 2(2): 270, 270pl, 271d,
    272
radians 2(2): 268
rectispinus 2(2): 254
  var. *robustus* 2(2): 249, 253
reichenbachii var. *chisoensis* 2(2): 261

rigidissimus 1(4): 170-175
  var. *rigidissimus* 1(4): 174-175
  var. *rubispinus* 1(4): 170pl, 173d,175
rufispinus 2(2): 266
russanthus 1(4): 169
salm-dyckianus 11(2): 79
scheerii 1(4): 150-5; 11(2): 80
  var. *gentryi* 1(4): 150pl, 151-152,
    153d, 154
schmollii 11(2): 71
schwarzii 2(2): 268
stoloniferus 2(2): 255; 11(2): 70
  var. *stoloniferus* 2(2): 256, 256pl, 257d
  var. *tayopensis* 2(2): 255, 256, 258
stramineus 1(4): 160-163, 160pl, 162d
'subinermis group' 11(2): 70-71
subinermis 2(2): 255; 11(2): 70-74, 70pl,
  72d, 79
  var. *aculeatus* 11(2): 73
  var. *ochoterenae* 11(2): 71-72, 74
  var. *subinermis* 11(2): 70-74
viereckii 1(4): 163-165, 164d, 164pl
viridiflorus 1(4): 165-166
  var. *cylindricus* 1(4): 170
  var. *davisii* 1(4): 186
weinbergii 2(2): 272
wislizenii 11(2): 79
Sect. *Costati* 11(2): 71
Sect. *Erecti* 11(2): 70
Sect. *Reichenbachii* 11(2): 70-71
Sect. *Wilcoxia* 11(2): 71
Echinacea intermedia 4(1): 34
  purpurea 4(1): 34-35
  serotina 4(1): 34
Echinopsis amoena 2(2): 272
  scheeri 11(2): 80
  valida var. *densa* 1(4): 157, 159
Echites illustris 9(3): 133
ecology, Alpine 1(1): 30-35;
Economic Botanists at Kew, Two
  Centuries of 10(3): 132-138
ECOS 10(3): 135
Ecuador, Plant-hunting in the

19

mountains of 6(4): 172-181
Edible Barringtonias 9(4): 164-182
Edmonds, Jennifer 15(3): 186-193
Edwards, Brigid 10(1): 18pl
Edwards, Peter J. 8(3): 109-112,
   112-118, 118-123, 140-146, 155-158
Edwards, Sydenham 12(4): 236
Edwardsia 14(4): 228
 toromiro 14(4): 224
Eggli, Urs 9(2): 59-63
Elaeodendron orientale 13(3): 216
Elaphoglossum 8(3): 118
 crinitum 8(3): 118-123, 118pl, 120d,
   122d
 hirtum var. *micans* 5(2): 86d
Encyclia citrina 2(1): 195
 mariae 2(1): 195-197, 196pl, 197d
 mooreana 12(4): 239
Endemic flora of St. Helena 15(1): 28
Endemism in Malesian flora 12(2): 95-110
 in Turkey 2(4): 363-365
 in St Helena 2(4): 369-379
Engelhardt, O. von 14(1): 41
Engler, H.A. 12(3): 157-161, 158p
Entada 14(4): 234
Eperua 14(4): 243-244
Epicattleya 'Marina' 2(1): 195
Epidendron stanfordianum 3(1): 41
Epidendrum ilense 8(1): 5-8, 6pl, 8d;
   11(1): 35
 Plastic Doll 8(1): 7
 mariae 2(1): 197
 melanocaulon 5(4): 156
 whartonianum 5(4): 156
Epigeneium 13(3): 139-145
 acuminatum 13(3): 141
 cymbidioides 13(3): 141
 fargesii 13(3): 140
 longirepens 13(3): 140
 lyonii 13(3): 141
 pulchellum 13(3): 140
 speculum 13(3): 141
 treacherianum 13(3): 141

 triflorum 13(3): 140-143
  key to varieties of 13(3): 142
  var. *elongatum* 13(3): 141-142
  var. *orientale* 13(3): 139-145, 140pl,
   144d
  var. *triflorum* 13(3): 141-142
Epimedium 12(1): 15-25
 New large-flowered Chinese species of
  10(4): 178-184
 New Chinese Taxa of 12(1):15-25
 Acuminatum Group 12(1): 16
 acuminatum 12(1): 22-23
 *alpinum*12(1): 21, 24
 Davidii Group 12(1): 16
 davidii 12(1): 16-17,19d, 21-22
 diphyllum 12(1): 23
 dolichostemon 12(1): 16
 fangii 12(1): 16-17, 19d, 20d, 21-23
 fargesii 12(1): 16
 flavum 12(1):16-17, 21, 22pl, 24
 grandiflorum 2(1): 227; 12(1): 23
 latisepalum 10(4): 180, 181d; 12(1): 16
 leptorrhizum 12(1): 24
 ogisui 10(4): 182, 183d; 12(1): 16
 omeiense 12(1): 16, 22
 *pauci*florum 12(1): 16,17d, 23-25
 platypetalum 12(1):16, 24
  var. *tenuis* 12(1): 23-24
 *pubigerum*12(1): 21, 24
 versicolor 'Sulphureum' 12(1): 22
  'Neosulphureum' 12(1): 22
 series *Brachycerae* 12(1): 16
 series *Dolichocerae* 12(1): 16
Epipactis 10(1): 18
 abortiva 10(4): 163
 mairei 10(1): 21d
 royleana 10(1): 18-22, 18pl, 21d
Epipogium aphyllum 3(1): 9d
Epipremnopsis 12(2): 85,87
Epipremnum 12(2): 87
Eranthis hyemalis 4(1): 35-36
Eria 2(2): 283
Erica andevalensis. Natural habitat

in SW Spain, 2(3): 324-330, 327d
  association with mines 2(3): 325-328
  distribution in S Spain 2(3): 324m
ciliaris 2(3): 318d
Erigeron oharai 4(3): 119-123, 120pl, 122d
Erodendrum aemulum 11(4): 176
Erythrina 14(4): 182
Erythronium dens-canis 4(1): facsimile t.5, 38
Erythrophleum couminga 14(4): 237
Erythrorhipsalis pilocarpa 14(3): 127
Eschscholtz, J.F. 14(1): 41
Espeletia schultzii 7(4): 165-169, 166pl, 168d
Estonia 14(1): 35-55
Eucomis schijffii 6(2): 51-54, 52d, 52pl
Euhaynaldia organensis 9(3): 122
Eulophia 13(2): 82-85
  emarginata 13(2): 84
  euglossa 13(2): 82
  gracilis 13(2): 83
  guineensis 13(2): 82
  macrostachya 13(2): 83-84
  papuana 13(2): 84
  pulchra 13(2): 82-85, 82pl, 85d
  rouxii 13(2): 84
  silvatica 13(2): 84
  striata 13(2): 84
Eulophidium pulchrum 13(2): 83-84
  silvaticum 13(2): 84
Euodia balansae 5(4): 185
  bodinieri 5(4): 187
  daniellii 5(4): 182
  delavayi 5(4): 182
  fargesii 5(4): 185
  fraxinifolia 5(4): 188
  glauca 5(4): 185
  henryi 5(4): 182
  hupehensis 5(4): 182
  impellucida 5(4): 188
  meliaefolia 5(4): 185

officinalis 5(4): 187
rugosa 5(4): 187
rutaecarpa 5(4): 187
subtrigonosperma 5(4): 188
sutchuenensis 5(4): 182
velutina 5(4): 182
vestita 5(4): 182
Euphorbia columnaris 4(2): 99
  francoisii 5(2): 88-92, 90d
    var. *francoisii* 5(2): 89
    var. *rakotozafyi* 5(2): 89
  schillingii 4(3): 110-113, 110pl, 112d; 8(1): 45; 14(3): 157
    var. *praecox* 8(1): 45, 46
  subgen. *Paralias* 4(3): 113
  subsect. *Purpuratae* 4(3): 113
European Garden Flora 1(3): 119-122
Evans, Sir Geoffrey 10(3): 132
Everest, Mount 14(3): 153-166
Everlasting flowers 13(4): 195-196
Exochorda giraldii 3(4): 156, 158
    var. *giraldii* 3(4): 157
    var. *wilsonii* 3(4): 156-158, 156pl, 158d
    racemosa var. *wilsonii* 3(4): 158
Exostyles 14(4): 194
Eyde, Richard 6(2): 72-83

Fabaceae 14(4): 179
Fabales 14(4): 179
Fabricia laevigata 4(1): 40; 12(4): 236
Fadyenia fadyenii 8(3): 146
  hookeri 8(3): 140-146, 140 pl, 145d
  prolifera 8(3): 146
Fadyenii hookeri, variation 8(3): 140-146
Fagus hayatae 12(1): 42-45, 43d
Fairy Slipper (*Calypso bulbosa*): 3(4): 147-151
Farjon, Aljos 11(4): 156-166
Farrer, Ann 4(3): 110pl, 4(4): 180pl, 188pl; 6(2): 52pl, 56pl, 6(4): 148pl; 7(1): 14pl, 18pl, 20pl, 7(3): 118pl, 124pl, 7(4): 166pl; 8(1): 2pl, 8(4): 160pl, 176pl, 186pl; 9(2): 68pl, 72pl;

10(1): 14pl, 10(3): 118pl, 119d; 11(1): 6pl, 8d, 60pl, 62pl; 11(3): 128pl; 11(4): 156pl; 12(1): 10pl, 12pl, 14pl; 12(2): 68pl, 86pl, 88d; 12(3): 128pl, 132pl, 136pl, 138pl, 142pl, 146pl; 13(2): 64pl; 13(3): 134pl; 13(4): 192pl, 196pl; 15(1): 2pl

Farrer, Reginald, and Glasnevin 10(4): 171-8

Fascicularia 13(1): 22

Faurie, Père Urbain 10(4): 185

Ferguson, Keith 6(2): 74-83; 13(4): 200-203

Fern herbarium of Thomas Moore, The 8(3): 147-154

Fernandez, Antonio 15(1): 47d

Ferns, fish-bone 8(3): 112

lace 8(3): 113

ladder 8(3): 112

rare Macarene Islands 13(4): 233

sword 8(3): 112

tassel 8(3): 126

Ferocactus glaucescens 11(2): 71

schwarzii 11(2): 71

Ferrer, Ricard 15(1): 69pl(b/w), 71pl(b/w)

Fici, S. 13(2): 105-106

Ficus 13(2): 105

columnaris 13(2): 105-106

macrophylla 13(2): 106

subsp. *columnaris* 13(2): 105-107, 106d

subsp. *macrophylla* 13(2): 106-107

magnolioides 13(2): 106

subsp. *magnolioides* 13(2): 105

var. *macrophylla* 13(2): 105-106

var. *magnolioides* 13(2): 105-107

nervosa 13(2): 105

Field, D.V. 10(3): 135

Field, Ruth 10(3): 144-148

Firmiana 11(2): 88

Fischer, F. von 14(1): 40

Fitch, J.N. 13(4): 232d

Fitch, Walter Hood 1(1): 44-45, 1(1), 1(2), 1(3), 1(4): Cover pls; 9(1): 4pl, 8pl,

10pl, 12pl, 16pl, 18pl; 11(4): 184pl, 187, 195, 12(4): 238

Fitzroya cupressoides 4(3): 131, 132, 132d

Flacourtia indica 13(4): 222

Flamboyant tree 14(4): 234

Flanagan, Mark 5(4): 181-191, 6(3): 134-139; 12(1): 25-28, 28-42, 42-45

Flickingeria 13(3): 141

Fliegner, Hans J. 8(2): 58-63

Flora Brasiliensis 13(3): 155

Flora of Turkey 2(4): 357-367

Flora Malesiana 12(2): 51-62, 73

Flora Orientalis 11(3): 134-136, 140, 144-145

Floribunda 13(2): 73

bahiensis 13(2): 73-74

pusilliflora 13(2): 73

Forrest, George 13(2): 108-109

Forster, J.R. & G. 14(4): 227

Forsythia suspensa 2(1): Cover pl

Foster, Clive 14(4): 194-197, 242-244

Fothergill, Mark 4(3): 120pl, 124pl, 4(4): 154pl, 158pl; 5(1): 8pl, 12pl, 16pl, 18pl, 5(2): 52pl, 60pl, 70pl; 5(4): 148pl; 9(3): 116pl, 120pl, 124pl, 128pl, 132pl, 134pl; 10(1): 60pl, 10(3): 112pl, 126pl, 128d, 130pl, 130d; 11(1): 14pl, 17d, 18pl, 21d, 27d; 11(2): 52pl, 54d, 90d; 11(3): 110d, 112pl, 114pl, 116pl; 12(1): 26pl, 27d, 12(2): 74pl, 76d, 78pl; 12(3): 128d, 129d, 133d, 136d, 139d, 143d, 146d; 13(1): 4pl, 6d, 22pl, 25d, 34d; 13(2): 82pl, 85d; 13(4): 204pl, 208d, 210pl; 14(1): 8pl, 11d; 14(2): 78pl; 14(4): 214pl; 15(4): 232pl

Foxia spicata 11(3): 128, 130

Frankenia portulacaefolia 2(4): 372-373

Franklinia alatamaha 15(2): 125-126

Friar's Cowls & Mouse Plants 7(1): 14-21

Friis, Ib 4(2): 97-101

Fritillaria 13(1): 27-32

bhutanica 9(2): 54

chitralensis 13(1): 27-32, 28pl, 31d
delavayi 9(2): 51-54, 53d, 54pl
eduardii 13(1): 28-29
imperialis 13(1): 27-29
  var. *chitralensis* 13(1): 27-29
raddeana 13(1): 28-29
  subg. *Petilium* 13(1): 28
Froebel, O. 14(2): 68, 72
Fryer, Jeanette 12(4): 198-201
Furneaux, Capt. Tobias 14(4): 222
Fynbos 13(2): 56-58

Galathea speciosa 9(1): 8
Galanthus elwesii 1(1): 25
Galeandra montana 13(3): 155
  viridis 13(3): 155
  spp. 13(3): 157
Garcinia livingstonei 5(3): 122d
Gardenia ternifolia var. *goetzei* 4(3): 142d
Gardner, David 15(1): 12-19; 15(4): 231-235
Gardner, Martin 3(3): 140-142; 4(3): 127-140
Gastrodia 1(2): 88-90
Genève, Conservatoire et Jardin Botaniques 11(3): 135, 141
Genoplesium 1(2): 89
Gentiana depressa 14(3): 158
  laksnakarae 1(2): 80, 81d
  ornata 5(1): 26d; 14(3): 164d
  triflora 2(1): 227
  urnula 1(1): 35, 35d
  verna 2(3): 314d
  'Clare' 2(3): 314
Geranium 8(1): 51-53
  cataractarum 8(1): 51-53, 53d, 52pl
    subsp. *pitardii* 8(1): 51
  farreri 1(3): 111
  napuligerum 1(3): 114
  orientalitibeticum 1(3): 112-114, 117-118, 117d, 118pl
  pylzowianum 1(3): 111-116, 115d, 116pl
  stapfianum 1(3): 112, 114

var. *roseum* 1(3): 112, 113
Gereau, Roy E. 8(4): 194-197
Germann, G.A. 14(1): 37
Gesneria sylvatica 12(4): 192
Gesneriaceae 11(2): 82; 12(4): 191-194; 15(1): 49, 62
Ghost Orchid (*Epipogium aphyllum*): 3(1): 9d
Gigantochloa 12(1): 9
Gladiolus aureus 1(1): 28, 29; 4(4): 199-201
Glasnevin 10(4): 171-177; 12(4): 182-185, 209-220; 15(4): 275
Glasnevin, artists 15(4): 279-280
Glasnevin, a medley of plants from 12(4): 236-244
Glasnevin, Curvilinear Range at 12(4): 209-220
Glasnevin, Kew and *Curtis's Botanical Magazine* 12(4): 182-185
Glasnevin Orchids 12(4): 237
Glaucidium palmatum 2(1): 228-229, 228pl
Glaucium flavum 6(4): 147-151, 148pl, 149d
  luteum 6(4): 151
  tricolor 6(4): 151
Gleditsia 14(4): 176, 181
Gliricidia 14(4): 176, 182
Gloxinia sylvatica 12(4): 191-194, 194d, 192pl
  'Glasnevin Jubilee' 12(4): 192
  tubiflora 12(4): 239
Glycine clandestina 3(4): 176, 177d
Glycyrrhiza 14(4): 182
Goaman, Victoria 5(4): 164pl, 166pl; 6(1): 4pl, 6pl, 6(2): 58pl, 6(3): 122pl; 7(3): 110pl, 7(4): 180; 8(2): 54pl; 11(3): 124pl
Goldblatt, Peter 11(1): 37-38
Golfballia ambusta 3(4): 181d
Goodia 11(4): 172
Gould, Elizabeth 11(4): 189-191

Gould, John 11(4): 183, 186-197, 188p
and Curtis's Botanical Magazine 11(4):
186-197
Goyder, David 7(1): 3-6, 7(4): 169-173;
9(3): 131-134
Gramineae 12(1): 8-15
Graphorkis 13(2): 82
blumeana 13(2): 84
macrostachya 13(2): 84
pulchra 13(2): 84
Gravereaux, J. 14(3): 64, 72
Gray, Asa 11(3): 140
Green, Peter S. 6(3): 116-124; 8(2): 58-63;
9(2): 63-7; 10(3): 113-116; 11(4): 183-
186
Greenop, Helen 13(3): 140pl, 144d; 14(3):
130pl, 134d
Green Pitcher Plant Recovery 10(3): 147
Grenier, J.C.M 14(2): 64
Grey-Wilson, Christine 10(3): 123d; 11(2):
62d, 64d; 14(1): 23-30
Grey-Wilson, Christopher 1(1): 30-35;
1(2): 49-50, 92-93; 1(3): 97-99; 1(4):
145-146, 2(1): 193-194, 201-207, 232-
234; 2(2): 241-242; 2(3): 289-290, 330-
331; 2(4): 337-338; 3(3): 99-102, 104-
109; 4(1): 3-6, 35-38, 4(3): 116-119;
5(1): 7-11; 6(4): 147-151; 8(4): 159-
162; 15(3): 180-185
Greyia sutherlandii 12(4): 240-241
Grierson, Mary 3(2): 78-82; 3(3): Cover pl,
106pl; 14(2): 82pl, 85d
Grimshaw, J. 14(1): 23-30; 15(1): 37-41
Gumwood (*Commidendrum*): 2(4): 374-
375
Gunnera macrophylla 2(2): 279
Gunpowder Plant (*Pilea microphylla*):
1(1): 17
Gurr, Linda 15(3): 170pl, 173d
Guttiferae 11(2): 2): 65; 15(2): 92
Guzmania multiflora 8(2): 63-66,
64pl, 66d
Gymnadenia conopsea 11(1): 24

Gymnogramme aurita var. *levingei* 12(4):
196
levingei 12(4): 197
Gymnomesium pictum 5(2): 76
Gynura auriformis 7(3): 112
brownii 7(3): 112
scandens 7(3): 109-113, 110pl,110d,
111d
taylori 7(3): 112

Haageocereus faustianus 15(2): 99
Habranthus martinezii 4(3): 107-110,
108pl, 109d
Haematoxylum campechianum 14(4): 191
Haenke, T. 14(2): 87-100
Halfordia 4(4): 173
Hall, Tony 1(2): 51; 12(1): 5-6; 13(1): 4;
14(3): 137-141; 15(4): 242-248
Halliday, Pat 1(1): 2, 20-22; 1(3): 108-111;
3(2): 51-55; 4(2): 59-61, 4(4): 165-168;
6(3): 110-112; 7(3): 109-113, 7(4): 176-
9; 9(1): 15-18; 11(3): 126d; 11(4): 177;
12(1): 43d; 12(4): 189d, 210d, 208d;
13(2): 64d; 13(2): 109d; 14(1): 33d;
14(3): 140d
Halliwell, Brian 2(1): 226-232
Hammersmith Iron Works, Dublin 12(4):
212d
Hardenbergia 14(4): 182
Harley, Raymond 3(4): 151-156; 9(3): 103-
116, 134-138
Harriott, Georita 14(3): 126pl, 128d
Hart-Davies, Christine 10(4): 162pl, 164pl;
11(1):22pl
Hartzer, Pascale 14(1): 16-22, 21d
Harvey, Yvette 9(1): 3-7, 9(3): 120-124
Harvey, William Henry 12(4): 202-203;
13(1): 36-41, 37p
Harveya pulchra 12(4): 202-206, 202pl
Hastingia coccinea 13(2): 80
Hatiora 14(3): 125
Hawaiian Mountain Plants, Some
5(2): 82-88

Hay, A. 14(2): 82-86
Haynaldia organensis 9(3): 122
Haywood, Helen 5(4): 154pl; 7(3): 128pl;
  8(3): 112pl, 118pl, 124pl, 128pl, 134pl,
  140pl; 9(4): 162pl; 10(4): 185d, 187d,
  188d, 189d; 11(4):168pl, 172pl, 173d;
  12(1): 6pl, 7d, 32d, 39d, 40d; 12(2):
  90d, 92d, 92pl, 94d
Heaton, Edmund 15(2): 104-108
Hebe colensoi 7(4): 189d
Heberden, Ernest 7(3): 131-133
Heberden, Thomas 7(3): 131-133
He-Cabbage (*Senecio leucadendron*):
  2(4): 377
Hedysarum vespertilionis 13(1): 19-20
Heine, H. 3(4): 167-176
Helichrysum 13(2): 57-58
 adenocarpum 12(1): 5
  var. *alpinum* 12(1): 6
 formosissimum 12(1): 5
 hoffmannianum 12(1): 6
 kilimanjari 12(1): 5
 meyerjohannis 12(1): 5-8, 6pl, 7d
 nandense 12(1): 5
 newii 12(1): 8
 proliferum 13(2): 59
 series *Polylepidea* 12(1): 5
 series *Holohelichrysum* 12(1): 5
Helleborine lingua 3(1): 25
 royleana 10(1): 20
Helleborus argutifolius 4(4): 156d
 chinensis 15(4): 246
 x *hybridus* 15(4): 242
 hyemalis 4(1): facsimile t.3
 niger 4(4): 156d
 x *nigercors* 4(4): 154, 155, 156d
  'Alabaster' 4(4): 154-157, 154pl, 156d
 thibetanus 15(4): 242-248, 242pl, 247d
 vesicarius 2(4): 366d
 viridis var. *thibetanus* 15(4): 246
 sect. *Dicarpon* 15(4): 243
 sect. *Helleborastrum* 15(4): 242-243
 Henry, Augustine 3(3): 136-140; 14(3):

  149; 14(4): 249
Henry, Gerard 6(3): 106-109
Hepatica nobilis var. *japonica* 2(1): 228
Hepper, F. Nigel 5(2): 69-72
Heptacodium jasminoides 7(3): 133-138,
  135d
Hermans, Johan 13(3): 120-123
Herre, Hans 11(1): 10,11
Heteropappus krylovianus 4(3): 123
 oharai 4(3): 123
Heterotropa fudsinoi 9(2): 70
 minamitanianum 9(2): 74
 yakusimensis 11(2): 63
Hexastylis speciosa 5(4): 151-153,
  152pl, 152d
Heyderia 12(2): 66-67
Heyerdahl, Thor 14(4): 223
Hibbertia saligna 5(4): 172d
Hibiscus 13(4): 210-214; 14(4): 228
 arnottianus 13(4): 210-211
 bernieri 13(4): 210
 boryanus 13(4): 210-211
 columnaris 13(4): 211-212
 denisonii 13(4): 210-211
 fragilis 13(4): 210-214, 210pl
 genevi 13(4): 210-212
 kahili 15(4): 236
 kokio 13(4): 210; 15(4): 236
  subsp. *kokio* 15(4): 237
  subsp. *saintjohnianus* 15(4): 236-241,
   236pl, 239d
 liliastrum 13(4): 210
 liliflorus 13(4): 210-211, 213, 233
 perrieri 13(4): 210
 roetae 15(4): 239
 rosa-sinensis 13(4): 210-211
 saintjohnianus 15(4): 239
 schizopetalus 13(4): 210-211
 storckii 13(4): 210-211
 tiliaceus 3(2): 80d; 13(4): 211
 waimeae 13(4): 210
 sect. *Lilibiscus* 13(4): 210-211; 15(4):
  237

Higgens, Sarah 15(2): 124-132
Hildegardia 11(2): 88-94
 Ankaranensi s 11(2): 92
 australiensis 11(2): 89, 90d, 91
 barteri 11(2): 93
 cubensis 11(2): 92
 erythrosiphon 11(2): 92
 gillettii 11(2): 93
 merrittii 11(2): 91
 migeodii 11(2): 93
 perrieri 11(2): 92
 populifolia 11(2): 91
 sundaica 11(2): 89, 91
Hill, David A. 7(4): 182-186
Hillier, J.M. 10(2): 89
Himantoglossum 11(1): 22
 Adriaticum 11(1): 29d, 30-31
 caprinum 3(1): Cover pl, 28-31, 28pl
  var. *calcaratum* 3(1): 31
  var. *heldreichi i* 3(1): 31
 hircinum 3(1): 28, 29; 11(1): 22-28, 22pl,
  27d, 31
  subsp. *Calcaratum* 3(1): 28-29, 31
  subsp. *caprinum* 3(1): 31
  var. *heldreichii* 3(1): 31
Hind, Nicholas 4(1): 34-35, 4(3):
  119-123; 5(2): 55-60; 7(3): 109-113,
  128-131, 7(4): 165-169; 8(1): 1-5; 9(4):
  153-156, 160-164; 12(1): 5-8; 13(1): 3-
  7; 13(2): 56-62; 13(4): 191-199
Hippeastrum puniceum 14(2): 93d, 94pl
HMS 'Adventure' 14(4): 222
HMS 'Endeavour' 14(2): 87-88
HMS 'Investigator' 14(2): 91, 97
HMS 'Resolution' 14(4): 222
Hoaxes, botanical 3(4): 178-185
Hofbauer, Joseph 15(3): 209-210
Hoffmannseggella bahiensis 9(3): 118
Hohenbergia strobilacea 8(2): 68
Hokkaido forest 2(1): 226
Holland, J.H. 10(2): 90, 91
Holmskioldia 13(2): 79-81
 rubra 13(2): 80

 sanguinea 13(2): 79-81, 80pl, 81d
Homaida 7(1): 17
Homalium 13(4): 185
Hooker, Sir Joseph Dalton 10(2): 89
Hooker, Sir William Jackson 10(2): 87-88;
  13(1): 38-40
Hornbeam (*Carpinus*): 3(3): 137-138
Hortus Amstelodamensis 4(2): 78-91
Howea belmoreana 13(2): 107
 forsteriana 13(2): 107
Howes, Dr F.N. 10(2): 90, 91-92
How many species of *Allium* are known?
  9(4): 180-182
Hoya coriacea 7(1): 6
 costantinii 7(1): 6
 javanica 7(1): 6
 multiflora 7(1): 3-6, 4pl, 5d
 reticulata 7(1): 6
Huernia macrocarpa 15(1): 2-11, 2pl
Hummingbird, Temminck's sapphire-wing
  11(4): 183, 194d, 195
Hummingbirds 11(4): 171, 191-197
Hunt, David 2(4): 339-341; 4(2): 64-66
Hunt, Peter F. 10(1): 36
Huperzia 8(3): 126
 pinifolia 8(2): 123-8, 124pl, 125d
Hutchinson, Dr J. 10(2): 90, 91
Huttum edule 9(4): 176
Hyacinthaceae 11(3): 128; 14(1): 31
Hyacinthus corymbosus 5(1): 7;
 gawleri 5(1): 7
 paucifolius 5(1): 7
 spicatus 11(3): 128-129
Hybrid Cornus at Underway in
  Somerset, The 7(4): 179-182
Hydrodea cryptantha 2(4): 371-372
Hymenaea courbaril 14(4): 191
Hymenodium crinitum 8(3): 122
Hymenophyllum flabellatum 5(3): 132d
Hyophorbe 13(4): 186-190
 amaricaulis 13(4): 186-189
 indica 13(4): 186, 188
 lagenicaulis 13(4): 186-190, 186pl, 189d

vaughnii 13(4): 186, 233
verschaffeltii 13(4): 186, 188
Hypericum 12(1): 35-38; 15(2): 95
  bucklei 11(2): 66
  buckleyi 11(2): 65-69, 66pl, 68d
  dowellianum 11(2): 65
  siamense 15(2): 93
  subsessile 15(2): 92-95, 92pl, 94d
  sect. *Ascyreia* 15(2): 92-93
  sect. *Myriandra* 11(2): 66
Hyphaene compressa 5(3): 124d

Ibáñez, V. 14(2): 87-100
ILDIS database of legumes 14(4): 189-190
Ilex centrochinensis 1(1): 47-48
  corallina 1(1): 47-48
  cyrtura 1(3): 100-3, 100pl, 102d
  fargesii 3(3): 127-128, 130,131,132
    subsp. *fargesii* 3(3): 131, 132
    subsp. *melanotricha* 3(3): 131, 133d,
      134
    var. *brevifolia* 3(3): 130, 133d, 134, 135
    var. *fargesii* 3(3): 128-129, 129d, 132-
      133, 133d, 134
    var. *megalophylla* 3(3): 132
    var. *parvifolia* 3(3): 130-131, 133d, 134
    var. *sclerophylla* 3(3): 129-130
  fargesii complex 3(3): 127-135
  fargesii sens. lat.: distribution of 3(3):
    127m
  franchetiana 3(3): 128, 132
    var. *parvifolia* 3(3): 130,134
  melanotricha 1(3): 100; 3(3): 131
Impatiens columbaria 15(1): 37-41, 38pl,
  40d
  digitata 14(1): 27
  flanaganae 2(4): 342-345, 342pl, 344d
  kazkazini 14(1): 23, 27, 29
  kilimanjari 14(1): 23-30
    subsp. *pocsii* 14(1): 23-30, 24pl, 28d
  lateritia 14(1): 26
  meruensis 14(1): 27
  niamniamensis 13(3): 173d, 174

  phuluangensis 1(2): 80, 80d
  pseudoviola 14(1): 23-29
  rubromaculata 14(1): 27
  simbiniensis 14(1): 227
  tinctoria 2(4): 343
  walleriana 2(4): 342
Indian pink 9(3): 135
Indigofera 14(4): 177, 182, 213-220, 237
  amblyantha 14(4): 215
  compressa 14(4): 237
  cylindrica 14(4): 213, 216
  decora 14(4): 215
  dielsiana 14(4): 215
  frutescens 14(4): 213, 216
  gerardiana 14(4): 215
  heterantha 14(4): 215
  himalayensis 14(4): 215
  incarnata 14(4): 215
  jucunda 14(4): 213-220, 214pl, 218d
  kirilowii 14(4): 215
  pendula 14(4): 215
  potaninii 14(4): 215
  pseudotinctoria 14(4): 215
  subverticillata 14(4): 215
  sect. *Psiloceratiae* 14(4): 213
Inga 14(4): 176, 179, 187-189, 192, 243-
  244
"In honour of Ireland" 4(1): 39-51
International Organisation for the Study of
  Succulent Plants (IOS): 1(4): 187
International Union for Conservation
  of Nature & Natural Resources
  (IUCN): 1(1): 23; 1(3): 98-99
Intsia 14(4): 234, 235
Iochroma grandiflora 2(4): Cover pl
Ioniris ternax 8(1): 10
IOS 1(4): 187
Ipomoea alatipes 11(3): 151
Ireland, Helen 11(4): 170d
Iridaceae 15(2): 120
Iris aitchisonii 8(4): 172-175, 172pl, 175d
  var. *chrysantha* 8(4): 172-174
  var. *flaviflora* 8(4): 174

alata 3(2): 65
aphylla 1(3): Cover pl
dolichosiphon 7(1): 9-13, 10pl, 11d
gormanii 8(1): 10
lortetii 1(1): 24
kamaonensis 5(1): 30d; 7(1): 13-14
kemaonensis 7(1): 14; 13(2): 98
  Note on nomenclature 7(1): 13-14
microptera 3(2): 65
persica 4(1): facsimile t.1, 11, 33-34
planifolia 3(2): 61-65, 62pl, 64d
  f. *alba* 3(2): 63
scorpioides 3(2): 65
section *Limniris*, series *Californicae*
  8(1): 8-10
subgen. *Scorpiris* 4(1): 33
tarhunensis 3(2): 65
tenax 8(1): 8-11, 10pl, 11d
  var. *gormanii* 8(1): 10
transtagana 3(2): 65
trialata 3(2): 65
winogradowii 1(1): 24; 2(3): 330-331,
  331d
Irish Lace 12(4): 220-236
Irvingia gabonensis 13(3): 168
  grandifolia 13(3): 168
Irvingiaceae 13(3): 168
Island floras, susceptibility to
change 2(4): 337-338
Island Lobelia (*Trimeris*
  scaevolifolia): 2(4): 377
Island Redwood
  (*Trochetiopsis erythroxylon*): 2(4): 374
IUCN 1(1): 23; 1(3): 98-99
Ixora debilis 15(1): 45
  Pavettaefolia 15(1): 45
  ternifolia 10(4): 169

Jacquin, N.J. von 15(3): 157, 194-197
Jaquin's 'Selectarum Stirpium
  Americanarum Historia' title pages 15(3):
  194-214, 198fig., 200pl, 202fig., 202pl,
  204pl, 206fig., 208fig.

Jade Vine (*Strongylodon macrobotrys*):
  1(4): 188-190, 189d
Japan, Plant hunting in the Nara
  Prefecture of Honshu 10(4): 184-190
Japanese pagoda tree 14(4): 222
Japanese species of *Asarum*, Two 9(2):
  67-75
Japonasarum caulescens 11(2): 61
Jasminum grandiflorum ('grandiflora'):
  9(2): 65
  officinale 'Inverleith' 9(2): 63-67, 64pl,
  66d
  sinense 10(3): 113-116, 114pl, 116d
Jatropha phyllacantha var. *repanda*
  2(1): 223;
Jay, Maurice 14(2): 64-76
Jebb, Matthew 9(4): 164-182
Jellett, Barbara 14(2): 64-76
João Barbosa Rodrigues, Orchid paintings
  of 13(3): 152-157
Johns, R.J. 8(3): 109-112, 123-128, 128-
  33, 133-140, 147-154; 12(2): 52-62, 66-
  72, 77-85, 89-95, 95-110
Johnson, Magnus 3(2): 73-78;
Johnson, Margaret 3(2): 68-72; 10(3):
  124-131
Jonorchis abortiva 10(4): 164
Jubba river 5(3): 120-129
Juglans regia 13(2): 101
Juncus spp. 13(2): 98
Juno aitchisonii 8(4): 174
Jupunba pseudo-tamarindus 14(4): 200
Jura, region of France 14(2): 65

Kakadu National Park 11(2): 89
Karaka (*Corynocarpus laevigatus*): 3(3):
  125-126, 125d
Karomia 13(2): 79
Katherinea 13(3): 140
Kennedia 14(4): 182
Kent, Douglas H. 5(1): 32-36
Kerchove, Omer van de 15(1): 42pl
Kew's Living Collection of Orchids, Role

of 11(1): 32-37

Khumbu, Nepal 14(3): 153-166

Kilimanjaro, Mt 14(1): 23

Kinabalu, Mt, Sabah, N. Borneo 2(2):
    273-284, 274m

King, Christabel 1(1): 2, 6pl, 16pl,
    20pl, 44-5; 1(3): 100pl, 104pl,
    106pl, 108pl, 116pl, 118pl; 1(4):
    146, 150pl, 156pl, 160pl, 164pl,
    166pl, 170pl; 2(1): 198pl, 208pl;
    2(2): 244pl, 256pl, 260pl, 262pl,
    266pl, 270pl; 2(4): 340pl, 342pl,
    354pl; 3(2): 62pl, 68pl; 3(3):
    100pl, 112pl, 116pl; 3(4): Cover
    pl, 148pl, 152pl, 156pl; 4(2):
    64pl, 4(4): 160pl, 166pl; 5(2):
    56pl, 66pl; 6(1): 12pl; (caption in
    error - see p.144): 6(3): 96pl,
    102pl, 106pl, 114pl, 6(4): 164pl;
    7(2): 62pl, 64pl, 68pl, 72pl, 78pl,
    82pl, 7(4): 162pl, 170pl, 174pl;
    8(2): 60pl, 61d; 8(4): 168pl, 172pl;
    9(2): 56pl, 64pl, 9(4): 156pl; 10(1):
    30pl, 10(2): 68pl, 70pl, 82pl, 10(3):
    114pl, 10(4): 156pl; 11(2): 66pl,
    68d, 70pl, 72d; 11(3): 104pl, 107d;
    11(4): 174pl; 12(1): 17d, 19d, 20d, 22pl;
    12(4): 208pl; 13(1): 8pl, 12d, 20pl, 21d;
    13(2): 56pl, 60d, 72pl, 74d, 76pl, 80pl,
    81d; 13(3): 124pl, 128d, 132d; 13(4):
    186pl, 189d; 14(1): 2pl, 5d, 32pl; 14(2):
    78pl, 80d, 100-101; 14(3): 120pl, 123d;
    14(4): 194pl, 196d, 198pl, 201d, 204pl,
    205d, 225d; 15(1): 12pl, 17d, 28pl, 34d;
    15(3): 158pl, 161d, 176pl, 178d, 180pl,
    184d; 15(4): 220pl, 224d, 229d, 236pl,
    239d

Kingdon Ward, The lost Dogwood of
    Frank 6(2): 74-83

Kingdon Ward, F. 13(2): 108

Kleinia saginata 6(4): 151-156, 152pl,
    154d

Kleinodoxa gabonensis 13(3): 168

Knees, Sabina 5(2): 76-82, 88-92; 6(3):
    106-109, 6(4): 163-166

Kolbeck, Jiri 13(1): 33-35

Kowhai 14(4): 222

Koya-Obako mountains, The 10(4): 185d

Krapfia weberbaueri 15(4): 234

Kuhlhassettia javanica 2(2): 280

Labiatae 11(2): 52; 11(3): 109; 13(2): 79
    subfam. *Scutellarioideae* 13(2): 80

Laburnum 14(4): 176

Lace, Irish 12(4): 220-236

Lachanodes arborea 2(4): 374d, 376-377

La Croix, Eric 13(3): 165m, 166m

La Croix, Isobyl F. 8(2): 78-81; 11(2):
    55-59, 82-88; 13(3): 146-152, 165-176,
    171d, 173d; 14(3): 166-169; 15(1): 20-
    26; 15(3): 169-174; 15(4): 249-254

l'Hay-les-Roses 14(2): 64, 68-69, 72

Lack, H.W. 14(2): 87-100; 15(3): 157,
    194-214; 15(4): 254-272

Lacoski, Rachel 14(2): 64pl

Lady's Slipper Orchids:
    (*Cypripedium calceolus*): 1(3): 139-42;
        14(3): 148
    (*Cypripedium japonicum*): 3(1): 44d

Laelia bahiensis 9(3): 116-120, 116pl,
    119d
    purpurata 13(3): 159, 160d

Lahe Rara, Nepal 13(2): 96, 98

Lambkin, Deborah 12(4): 192pl, 193d

Lambourne, M. 11(4): 186-197

Lammers, Thomas G. 10(2): 70-75

Lamond, Jennifer 1(1): 2, 9-11; 7(3): 121-
    124

Lancaster, R. 13(2): 108

Langhorne, Joanna 1(2): 50, 58pl, 62pl;
    3(3): 110pl; 4(3): 114pl, 116pl;
    5(4): 152pl, 158pl; 6(1): 16pl; 8(1):
    10pl; 13(1): 28pl, 31d; 14(3): 142pl,
    146d; 14(4): 208pl, 211d; 15(4): 242pl

Lantana 13(2): 80

Lastrea bukoensis 12(4): 197

29

levingei 12(4): 197
Latania 13(4): 186
Latourorchis alexandrae 7(3): 120
Laurel 1(3): 133-139
  Californian (*Umbellularia californica*):
  1(3): 133
  Canary (*Laurus azorica*): 1(3): 135
  Mediterranean Bay (*Laurus nobilis*): 1(3):
  133
Laureola 4(4): 176
  fragrans 4(4): 189
Laurus nobilis 1(3): 133
Lavrih, Cherry-Anne 10(1): 22pl
Lawai Kai, Hawaii: Allerton Garden 3(2):
  78-82, 79d, 81d, 82d
Leach, G. 11(2): 88-94
Lear, Michael 2(1): 201-204
Lecanopteris 12(2): 89-95
  lomarioides 12(2): 89-95, 90d, 92d, 92pl,
  94d
  sarcopus 12(2): 91, 93
Lecointea 14(4): 194
Ledebour, C.F. von 14(1): 35-55, 39p
Legume activities at Kew 14(4): 184-193
Legumes in the Caatinga, Brazil
  2(1): 214-215, 215m, 217d, 220d, 222d,
  224d
Legume genera of Madagascar 14(4): 240
Legume publications of RBG Kew 14(4):
  245
Legume research at Kew 14(4): 186-192
Legumes in streets and gardens 14(4): 176-
  183
Legumes, Herbarium collections at Kew
  14(4): 184-185
Legumes, Living collections at Kew 14(4):
  184-186
Leguminosae 8(1): 54, 8(4): 197; 11(4):
  167, 171; 13(1): 19-21; 14(4): 175-250
  introduction to 14(4): 176-183
  of Madagascar 14(4): 231-241
  subfam. *Caesalpinioideae* 14(4): 179,
  180, 181, 187, 234

subfam. *Faboideae* 14(4): 179
subfam. *Mimosoideae* 14(4): 179, 187,
  198, 234
subfam. *Papilionoideae* 13(1): 19; 14(4):
  179, 180, 187, 194, 203-220, 234
tribe *Acaciaeae* 14(4): 187
tribe *Amherstieae* 14(4): 181
tribe *Amorpeae* 14(4): 182
tribe *Bossiaeeae* 14(4): 182
tribe *Caesalpinieae* 14(4): 180-182, 186,
  187
tribe *Carmichaelieae* 14(4): 182
tribe *Cercideae* 14(4): 180
tribe *Cicereae* 14(4): 180, 181
tribe *Coronilleae* 14(4): 180
tribe *Dalbergieae* 14(4): 182, 186, 187
tribe *Desmodieae* 14(4): 180, 182, 203
tribe *Detarieae* 14(4): 181, 186, 187, 192
tribe *Galegeae* 14(4): 180-182
tribe *Genisteae* 14(4): 180, 182
tribe *Hedysareae* 14(4): 181
tribe *Indigoferae* 14(4): 180, 182, 187,
  234
tribe *Ingeae* 14(4): 187, 199
tribe *Loteae* 14(4): 180, 182
tribe *Macrolobieae* 14(4): 181, 186, 187
tribe *Millettieae* 14(4): 182
tribe *Mimoseae* 14(4): 186
tribe *Mirbelieae* 14(4): 182
tribe *Phaseoleae* 14(4): 180-182
tribe *Podalyrieae* 14(4): 182, 187
tribe *Robinieae* 14(4): 180, 182
tribe *Sophoreae* 14(4): 182, 186, 187,
  192, 221
tribe *Swartzieae* 14(4): 186, 187, 194
tribe *Thermopsideae* 14(4): 180, 182
tribe *Trifolieae* 14(4): 180, 181
tribe *Vicieae* 14(4): 180, 181
subtribe *Bauhiniinae* 11(4): 167
subtribe *Bossiaeeae* 11(4): 172
subtribe *Cercideae* 11(4): 167
subtribe *Lespedezinae* 14(4): 203
subtribe *Cercidinae* 11(4): 167

le Lièvre, Audrey 5(4): 167-180; 7(4):
    186-200; 11(3): 131-143; 14(1): 35-55
Lemboglossum cervantesii 5(2): 51-55,
    52pl, 54d
Lemuropisum 14(4): 235
  edule 14(4): 235
Leontochir ovallei 14(1): 7-12, 8pl, 11d
Lepidopogon gnaphalodes 13(4): 197
  ponae 13(4): 197
Lepismium 14(3): 125
Leptocentrum spiculatum 4(2): 73
Leptogramme aurita 12(4): 197
Leptospermum 12(4): 186
  lanigerum 12(4): 186-190, 186pl, 189d
  parviflorum 12(4): 186
  recurvum 12(4): 186
  scoparium 12(4): 186
Lequeetia abortiva 10(4): 164
Lespedeza 14(4): 181, 203
  polyantha 14(4): 204
Leucadendron aureum 11(4): 175,176
Leucaena 14(4): 176
Leucojum fontianum 9(3): 159
  tingitanum 9(3): 156-160, 156pl, 159d
Leucomphalos 14(4): 234
Levinge, Henry Corbyn 12(4): 195-196
Lewis, Gwilym 1(4): 146, 188-189; 2(1):
    214-215; 2(4): 380-382; 9(1): 12-14;
    14(4): 194-197
Libocedrus 12(2): 66-67
  papuana 12(2): 70-71
  torricellensis 12(2): 71
  wigmannii 12(2): 70
Lièvre, Audrey le 14(1): 35-55
Ligustrum sempervirens 8(2): 59-62,
    60pl, 61d
Liliaceae 11(1): 18; 13(1): 27-32
Lilacs, The Laciniate-leaved 6(3): 116-124
Lilium 11(1): 18, 20
Lilium bulbiferum ssp. *croceum* 10(4):
    190-197, 191d
  subsp. *croceum*, Distribution maps 10(4):
    192d, 194d, 195d

  subsp. *croceum* in the Netherlands and
    Northern Germany 10(4): 190-197
  carniolicum 1(1): 43
  chalcedonicum 1(1): 36
  ciliatum 1(1): 38
  henryi 3(3): 137d
  lancifolium 2(1): 231
  martagon 1(1): 36
  medeoloides 2(1): 231, 231d
  pitkinense 1(1): 27
  pomponium 1(1): 36-37
  ponticum 1(1): 38
  pyrenaicum 1(1): 36-43, 39d
    subsp. *carniolicum* 1(1): 38, 39-40,
      41m, 43
    var. *albanicum* 1(1): 38, 40
    var. *bosniacum* 1(1): 38, 40
    var. *carniolicum* 1(1): 38, 40
    var. *jankae* 1(1): 38, 40
    subsp. *ponticum* 1(1): 37, 38, 42, 43
    var. *artvinense* 1(1): 38, 42
    var. *ponticum* 1(1): 38, 42
    subsp. *pyrenaicum* 1(1): 38-39, 43
    f. *rubrum* 1(1): 39
  tigrinum 2(1): 231
Limodorum 1(2): 88-90
  abortivum 10(4): 161-166, 162pl, 164pl,
    164d
    var. *anatolicum* 10(4): 164
    var. *rubrum* 10(4): 164
  pulchrum 13(2): 82-83
  trabutianum 10(4): 162
Limonia laureola DC. 4(4): 169, 170, 189
  laureola sensu Wallich 4(4): 191
Limosella aquatica 7(1): 36d
Lindera benzoin 1(3): 136
  umbellata 1(3): 135
Lindheimera texana 7(3): 128-131, 128pl,
    130d
Lindley, John 10(1): 36; 11(2): 76-77
Lindsaea repens 13(4): 233
Linnell, Ann 15(2): 138pl
Lippia 13(2): 80

Liquidambar orientalis 2(4): 361-362, 361d
Lissochilus pulcher 13(2): 84
Living Collections Division, Kew 3(1): 39-48
Livingstone, David 14(3): 166
Lizard Orchid (*Himantoglossum*): 3(1): 28-31
Lobelia 13(4): 200-203
  andropogon 9(1): 5
  blanda 10(2): 73
  bonplandiana 9(1): 5
  bridgesii 10(2): 70-75, 70pl, 74d
  cornuta 9(1): 4
  dentata 9(1): 5
  imperialis var. *kanitzii* 9(3): 122
  filiformis 13(4): 200
  laevigata 9(1): 5
  lucaeana 10(2): 73
  obscura 9(1): 4
  organensis 9(3): 120-124, 120pl, 123d
  purpurea 9(1): 5
  serpens 13(4): 200
  spectabilis 9(1): 5
  sphaerocarpa 9(1): 5
  surinamensis 9(1): 4
  thapsoidea 9(3): 122
  trapa 5(4): 161
  tupa 5(2): 84; 5(4): 157-161, 158pl, 159d
  vagans 13(4): 200-203, 200pl, 202d
  subg. *Lagotus* 13(4): 201
  sect. *Hemipogon* 13(4): 201
    subsect. *Leiospermae* 13(4): 201
Lodoicea 13(4): 222
Lomariopsis mauritiensis 13(4): 233
  variabilis 13(4): 233
Lomenia borbonica 11(1): 37
Lonchitis oxyglottis 3(1): 25
Lonchocarpus 14(4): 182, 189, 242-244
Lonicera pilosa 10(3): 111-113, 112pl, 113d
Loroglossum caprinum 3(1): 31
  var. *calcaratum* 3(1):31

  hircinum 11(1): 26
    var. *calcaratum* 3(1): 31
    subsp. *caprinum* 3(1): 31
    var. *caprinum* 3(1): 31
Lotus jacobaeus 4(2): 85d
Lourea 13(1): 19
  vespertilionis 13(1): 20
Lovett, Jon C. 3(2): 83-87; 6(1): 38-39, 6(2): 83-84; 8(4): 194-197
Loxanthocereus acanthurus 15(2): 99
  acanthurus var. *ferox* 15(2): 99
  bicolor 15(2): 100
  canetensis 15(2): 99
  cantaensis 15(2): 99
  convergens 15(2): 100
  erigens 15(2): 99
  erimiticus 15(2): 100
  eriotrichus 15(2): 99
  eulalianus 15(2): 99
  faustianus 15(2): 99
  gracilispinus 15(2): 99
  multifloccosus 15(2): 99
  neglectus 15(2): 100
  pacaranensis 15(2): 100
  peculiaris 15(2): 99
  pullatus 15(2): 99
    var. *brevispinus* 15(2): 100
    var. *chimbotensis* 15(2): 100
    var. *fulviceps* 15(2): 99
  xylorhizus 15(2): 100
Ludisia discolor 13(3): 161
Luetzelburgia auriculata 2(1): 222, 222d
Lupinus 14(4): 181
  arboreus 14(4): 192
Luzula wahlenbergii 6(1): 33d
Lycaste aromatica 11(1): 36
  brevispatha 10(1): 23, 24, 25
  candida 10(1): 22-27, 22pl, 26d
  horichii 1(3): 105
  tricolor 1(3): 103-105, 104d, 104pl
Lychnis bungeana 14(1): 53d
Lycopodium piscium 8(3): 127
  vriesianum var. *majus* 8(3): 126

32

Lysimachia minoricensis 15(2): 127
Lysionotus angustifolius 8(1): 22

Maackia 14(4): 182
Mabberly, D.J. 3(4): 167-176
Mackinder, Barbara 6(1): 6-8; 14(4): 221-226
Macnee, Sir Daniel 13(1): 36-40
Madagascar 13(2): 86-95
Madagascar Orchid Project 13(3): 163-164
Magnolia 'Albatross' 2(1): 201, 202d, 203pl
  campbellii 2(1): 203
  cylindrica 2(1): 201, 203, 204
  heptapeta 2(1): 203
  quinquepeta 4(1): 13d
  x *veitchii* 2(1): 203, 204
    'Peter Veitch' 2(1): 203
  wilsonii 4(1): 46d
Mahogany, Chinese 15(3): 188
Mahonia confusa 10(4): 156-161, 156pl, 159d
  confusa var. *bournei* 10(4): 160
  eurybracteata 10(4): 160
  fortunei 10(4): 156
  ganpinensis 10(4): 160
  russellii 6(2): 55-58, 56pl, 56d
  zemanii 10(4): 160
Maingaya malayana 10(2): 81-84, 82pl, 83d
Majorana tomentosa 11(3): 113
Malaxis punctata 4(2): 74-77, 74pl, 76d
Malawi, Streptocarpus in 11(2): 82-88
Malesia, an introduction 12(2): 52-62
  geological history 12(2); 59
  flora, endemism in 12(2): 95-110
  Dipterocarpaceae 12(2): 105
  Ericaceae 12(2): 105
  Gymnosperms 12(2): 103
  Pteridophytes 12(2): 102-103
Malvaceae 11(3): 146; 13(4): 210-214; 15(4): 236
Mammillaria fasciculata 2(2): 252

seemannii 11(2): 79
Mandevilla illustris 9(3): 131-134, 132pl, 133d
Manzanita, bigberry 15(3): 164
Maps
  Barun Khola, Nepal 1(1): 31
  Burren, position of 2(3): 312
  Cape York Peninsula 1(3): 123
  Distribution of *Erica andevalensis* in S Spain 2(3): 324
  Distribution of varieties of *Lilium pyrenaicum* subsp. *carniolicum* 1(1): 41
  Mt Kinabalu, Sabah, N. Borneo 2(2): 274
  Oaks, distribution in Turkey 2(4): 362
  Phu Luang Wildlife Sanctuary, Thailand 1(2): 76
  Quercus, distribution in Turkey 2(4): 362
  St Helena 2(4): 370
  São Raimondo Nonato, Brazil, Fundacão Ruralista, position of, 2(1): 215
  Turkey, 3 phytogeographical regions 2(4): 359
Marais, Wessel 1(3): 99, 106-107; 2(4): 368; 11(1): 37-38
Marcgraviaceae 14(3): 130-136
Marcgravia 14(3): 130-136
  comosa 14(3): 133
  myriostigma 14(3): 133
  nepenthoides 14(3): 132
  polyantha 14(3): 130-136
  umbellata 14(3): 130-136, 130pl, 134d
  bat pollination of 14(3): 132
Margorana dictamnus 11(3): 113
Marianne North 7(3): 138-142
  Gallery 13(3): 157-163
  Orchid Paintings of 13(3): 157-163
Marica caerulea 9(1): 8
  sabini 9(1): 8
Marojejya darianii 7(2): 90-95, 93d
Marshallia grandiflora 15(3): 158-163,

158pl, 161d
Marx, Roberto Burle 12(3): 147-150, 148p
Mascarena lagenicaulis 13(4): 188
 revaughnii 13(4): 188
Mascarene Islands 13(4): 181-237, 183m
Masdevallia chimaera 13(3): 120
 Chimaera Group 13(3): 120
 vampira 13(3): 120-121
 subg. *Lagotus* 13(3): 120
 sect. *Saccilabiatae* 13(3): 120
Mathew, Brian 1(2): 50, 68-74; 2(3):
 309-11; 3(1): 61-5; 3(4): 186-189
 4(1): 33-34, 38-39, 4(3): 107-110,
 4(4): 154-157; 5(1): 5-7, 5(2): 65-69;
 6(2): 51-54, 6(3): 112-115; 7(1): 13-
 14; 8(1): 8-11, 8(4): 172-175; 9(1):
 7-9, 9(4): 156-160; 10(3): 124-131;
 11(1): 3-5, 37-40; 11(3): 128-131;
 13(1): 27-32; 14(1): 31-34; 14(3): 137-
 153; 15(2): 104-108, 120-124; 15(3):
 164-168; 15(4): 242-248
Matonia, Matoniaceae 12(2): 96, 99-100
Matthews, Victoria 1(1): 2, 36-43;
 5(4): 157-161; 9(1): 10-12; 11(1): 18-
 22; 15(3): 175-179
Maund, Benjamin 15(4): 277
Maunder, Michael 5(3): 120-129, 137-140;
 10(4): 184-190; 14(4): 226-231; 15(2):
 124-132, 141-146
Mauritius 13(4): 182-185, 214-215, 217-
 233
Maw, George 3(4): 186-189, 187d, 188pl
Maxillaria tricolor 1(3): 105
Maximowicz, C. 14(1): 40
Mayo, Simon 1(2): 50, 51-67; 2(4):348-
 350; 3(3): 109-112; 4(2): 67-68, 4(3):
 114-116; 9(3): 124-127; 12(3): 147-150,
 150-153, 153-168
Mayombe 13(3): 165
McClintock, David 2(3): 324-330
Meadia hendersonii 14(3): 140
Meconopsis villosa 1(4): Cover pl
Medusagyne oppositifolia 6(4): 166-171,

169d, 170pl
Meikle, Desmond 9(1): 27-30; 11(3): 143-
 146; 12(4): 207-209
Melanthium gramineum 15(1): 70
 punctatum 15(1): 70
Melastomaceae 14(3): 169
Melhania melanoxylon sensu Melliss
 15(1): 33
Meliaceae 13(4): 212; 15(3): 186
Melianthus villosus 4(3): 123-127, 124pl,
 125d, 126d
Melocactus 9(3): 138; 13(2): 71
 conoideus 9(3): 138-142, 140d
 levitestatus 13(2): 76
Melville, R. 10(2): 91
Mesembryanthemaceae 14(1): 16
 subfam. *Ruschioideae* 14(1): 16
Mesembryanthemum intervallare 14(1): 20
Metalasia chimaera 13(2): 57
Metasequoia glyptostroboides 15(1): 79
Metrosideros collina
 var. *polymorpha* 5(2): 83d
Mexican orange 15(3): 175
Mexipedium xerophyticum 13(3): 130-133,
 130pl, 132d; 14(1): 60
Meyer, C.A. 14(1): 35-55
Micranthocereus 13(2): 71
Microcasia elliptica 12(3): 132
 muluensis 12(3): 132
 oblanceolata 12(3): 132
 pygmaea 12(3): 132
Microgramma 12(2): 90
Micropropagation Laboratory, Kew 11(1):
 32
Millettia 14(4): 182, 234
Miltonia candida 12(4): 238
Mimetes chrysanthus 11(4): 178pl, 178-
 183, 182d
 cucullatus 11(4): 179
 hirtus 11(4): 179
 palustris 11(4): 179
Mimosa 14(4): 177, 179, 228
Mimosaceae 14(4): 179

Minter, Sue 7(1): 22-31
Mioptrila odorata 15(3): 189
Mirbelia platylobioides 5(4): 177d
Misodendrum oblongifolium 4(3): 138, 139d
Mitchell, Robert B. 6(3): 124-134
Mitraria coccinea 4(3): 134, 135d
Moffett, Rodney 10(2): 64-66
Molpadia 13(4): 195
Monarrhenus 13(4): 196
Money Plant (*Pilea peperomioides*):
  1(1): 14-19, 16d, 16pl; 1(4): 144
Monkey Orchid (*Orchis simia*): 3(1): 34-6, 34pl, 35d
Monograph of the Trochilidae 11(4): 187, 193, 197
Moore, David 12(4): 237-240
  Frederick William 12(4): 238-242
  Lady Phyllis 10(4): 176
Moorea irrorata 12(4): 237
Moraceae 13(2): 106
Moraea alata 15(2): 106
  callista 8(4): 194-197, 195d
  loubseri 1(1): 28, 28d, 29
Morinda citrifolia 15(1): 42
Morley, Brian D. 12(4): 191-194
Mosquito bills 14(3): 139
Mossia intervallaris 14(1): 16-22, 16pl, 21d
Moths, pollination by 14(1): 17
Mottram, Roy 15(2): 96-103
Mucuna 14(4): 176, 182
  sempervirens 3(3): 139-140, 139d
Mudwort 7(1): 36d
Muelleranthus 11(4): 172
Muir, Harriet 1(3): 99, 139-140; 3(1): 8-13
Mundulea 14(4): 234
Murucuja millissima 11(4): 185
Mussaenda pavettaefolia 15(1): 45
Myosotidium hortensia 2(2): Cover pl
Myristicaceae 13(3): 168
Myrmecophila 12(2): 90-91
  lomarioides 12(2): 93

sarcopus 12(2): 93
Myrmecopteris 12(2): 90-92
  lomarioides 12(2): 93
  sarcopus 12(2): 93
Myrtaceae 12(4): 186-190
Myrtus communis, A forgotten cultivar of 9(1): 30-35
  'Baetica' 9(1): 31d
  var. *baetica* 9(1): 32-34

Narcissus cyclamineus 4(1): 43d
  minor 4(1): facsimile t.6, 38-39
Nascimento, Dulce 9(4): 154pl
  10(4): 168pl, 168d
Nastus borbonicus 13(4): 216
National Botanic Gardens, Glasnevin
  12(4): 182-185; 15(4): 275
National Collections, Wakehurst Place
  12(1): 28-42
Native orchids in archeological sites
  in Rome 10(1): 43-46
NCCPG 12(1): 28-42; 15(2): 85, 86-91, 136, 137
Nelson, E. Charles 2(3): 312-320, 324-330;
  3(3): 136-140; 3(4): 178-185; 4(1): 39-
  51; 5(1): 39-43, 5(3): 129-136; 7(1): 31-
  47; 8(1): 12-15, 8(2): 51-53, 81-89;
  8(4): 175-185; 10(4): 171-178; 12(4):
  182-185; 191-194, 195-197, 198-201,
  202-206, 220-236; 13(1): 36-41; 15(4):
  274-282
Neoglaziovia variegata 9(3): 107, 124-127, 124pl, 127d
Neolitsea sericea 1(3): 137, 138d
Neomarica caerulea 9(1): 7-9, 8pl
Neomoorea wallisii 12(4): 238
Neottia abortiva 10(4): 164
Nepal 1(1): 30-35, 31m; 13(2): 96-104;
  14(3): 153-166
Nepenthaceae 12(2): 111-118
Nepenthes 8(2): 89-90; 12(2): 111-118
  mirabilis 1(3): 131d
  rajah 2(2): 280, 284; 8(2): 82-94; 12(2):

Oncidium cervantesii 5(2): 55
  crispum 13(3): 153d (as *Rodriguezia eleutherosepala*)
  hastilabium 13(3): 136
  membranaceum 5(2): 55
  trilobum 13(3): 136
Onion, Paradoxical 4(4): 194-199
Ononis natrix 14(4): 191
Operculina pteripes 11(3): 151
Ophrys argolica 3(1): 14-16, 14pl, 15d
  cornuta subsp. *heldreichii* 3(1): 18
  cretica 3(1): 21-22
  doerfleri 3(1): 20-22, 20pl
  ferrum-equinum subsp. *spruneri* 2(4): 348
    subsp. *argolica* 3(1): 16
  galactosticos 3(1): 22
  galilaea 3(1): 20
  heldreichii 3(1): 18
  hiulca 2(4): 348
  kotschyi 3(1): 22
  lutea 3(1): 19
    f. *sicula* 3(1): 20
    subsp. *galilaea* 3(1): 20
    subsp. *murbeckii* 3(1): 20
    var. *lutea* 3(1): 19
    var. *minor* 3(1): 19-20, 20pl
  mammosa 2(4): 345
  murbeckii 3(1): 20
  oestrifera 3(1): 16
    subsp. *heldreichii* 3(1): 14pl, 14-15, 16-18, 17d
  scolopax subsp. *heldreichii* 3(1): 18
  sicula 3(1): 20
  sphegodes subsp. *spruneri* 2(4): 348
  spruneri 2(4): 345-348, 346pl, 347d
    f. *cretica* 3(1): 21
    f. *syriaca* 3(1): 21
    subsp. *cretica* 3(1): 21
Orange blossom, Mexican 15(3): 175
Orange lily 10(4): 190
Orchid distributions in Malawi, Some unusual 8(2): 78-81
Orchid-hunting in Choiseul 6(3): 124-134

Orchid Taxonomy at Kew 10(1): 36-38
Orchid painting from Palazzo Vecchio in Florence 11(1): 28-31
Orchid paintings of João Barbosa Rodrigues 13(3): 152-157
Orchid survey in Congo 13(3): 165-176
Orchidaceae 11(1): 22; 11(2): 55; 12(2): 62-66; 13(2): 82-85; 13(3): 120, 124, 130, 134, 139, 146; 14(1): 60; 14(3): 148; 15(1): 20; 15(3): 169; 15(4): 249
  subtribe *Dendrobiinae* 13(3): 140
Orchids, and Glasnevin 12(4): 237-239
  European: cultivation 3(1): 8-13
  Slipper (*Paphiopedilum*): of New Guinea and Solomon Islands 3(4): 159-166
  species in cultivation 3(1): 14-38
  underground 1(2): 84-91
  wild: exploiting 3(1): 41
  world crisis 3(1): 4-8
Orchis 11(1): 23
  abortiva 10(4): 163
  anatolica 3(1): 32-33, 32pl
    subsp. *sitiaca* 3(1): 32
  boryi 3(1): 34pl, 36-38, 37d
  caprina 3(1): 31
  coriophora subsp. *sancta* 10(1): 11
  coriophora var. *sancta* 10(1): 11
  hircina 11(1): 26
  laxiflora 10(1): 41
  lingua 3(1): 25
  macra 3(1): 36
  morio 10(1): 41, 44
    subsp. *boryi* 3(1): 38
  sancta 10(1): 9-12, 10pl, 12d
  simia 3(1): 34-36, 34pl, 35d
    var. *laxiflora* 3(1): 36
    var. *macra* 3(1): 36
  tephrosanthos 3(1): 36
  zoophora 3(1): 36
Origanum, Three membranous-bracted species of 11(3): 109-117
Origanum 11(3): 109-117
  acutidens 11(3): 111, 115-116

amanum 11(3): 110d, 114pl, 112-115
barglyi 11(3): 114
boissieri 11(3): 112
brevidens 11(3): 114
calcaratum 11(3): 111-112, 116
dictamnus 11(3): 110d, 111, 112pl, 112-113
haussknechtii 11(3): 116
laevigatum 11(3): 114
leptocladum 11(3): 111
libanoticum 11(3): 111
majorana 11(3): 109
microphyllum 11(3): 109
onites 11(3): 109
pseudodictamnus 11(3): 113
pulchellum 11(3): 112
rotundifolium 11(3): 110d, 111, 116pl, 115-117
saxatile 11(3): 113
scabrum 11(3): 111, 116
sipyleum 11(3): 111-112
syriacum 11(3): 109
vulgare 11(3): 109
x *hybridinum* 11(3): 112
'Barbara Tingey' 11(3): 116
'Kent Beauty' 11(3): 116
sect. *Amaracus* 11(3): 111
sect. *Anatolicon* 11(3): 111
sect. *Brevifilamentum* 11(3): 111, 114-116
sect. *Longitubus* 11(3): 111, 113
Ormindo, Paulo 13(3): 120pl, 122d
Orthophytum albopictum 2(4): 354-357, 354pl, 356d
burle-marxii 2(4): 357
Osteospermum sanctae-helenae 2(4): 371
Otacanthus coeruleus 8(4): 163-167, 164pl, 166d
Oxalis 10(2): 59
acutiuscula 10(2): 63
bachmannii 10(2): 63
brevicaulis 10(2): 63
canescens 10(2): 63

corniculata 10(1): 59
crassipes 10(2): 63
fortipes 10(2): 63
fulgida 10(2): 63
hirta 10(2): 59-63, 60pl, 62d
var. *canescens* 10(2): 63
var. *fulgida* 10(2):63
var. *hirta* 10(2): 63
var. *intermedia* 10(2): 63
var. *polioeides* 10(2): 63
var. *rubella* 10(2): 63
var. *secunda* 10(2): 63
var. *tenuicaulis* 10(2):63
var. *tubiflora* 10(2):63
hirtella 10(2): 63
latifolia 10(2): 59
macrostylis 10(2): 63
megalorrhiza 10(2): 63
multiflora 10(2): 63
pes-caprae 10(2): 59
rosacea 10(2): 63
rubella 10(2): 63
secunda 10(2): 63
tubiflora 10(2): 63
var. *canescens* 10(2): 63
var. *minor* 10(2): 63
var. *robusta* 10(2): 63
var. *secunda* 10(2): 63
Oxford Botanic Garden: Waterlilies 3(4): 167-176
Oxygraphis polypetala 13(2): 98
Oxypetalum caeruleum 12(4): 237
Oxytropis campestris 14(4): 192
halleri 14(4): 192
Pachypodium rosulatum var. *gracilius* 7(4): 169-73, 170pl, 172d
Pachysandra terminalis 2(1): 226
Paederia bojerana ssp. *foetens* 4(3): 143d
Paeonia cambessedesii 12(4): 242
emodii 13(2): 101
obovata 2(1): 229
Paepalanthus polyanthus 9(3): 111d
Palermo Botanical Garden 13(2): 105,

106d
Palm, big leaf 7(2): 90-95
Palm House at Kew, Replanting
    the 7(1): 22-31
Palmae 13(4): 186
  subfam. *Arecoideae* 13(4): 187
  subfam. *Ceroxyloideae* 13(4): 187
  tribe *Areceae* 13(4): 187
  tribe *hypophorbeae* 13(4): 187
Pamplemousses Botanic Garden 13(4):
    219, 226-227
Pandanus 13(4): 214, 216
  heterophylla 13(4): 216
Panicum issachnoides 5(2): 87d
Papaveraceae 15(1): 12
Paphiopedilum of New Guinea and
    Solomon Islands 3(4): 159-166
Paphiopedilum 13(3): 130
  bacanum 12(2): 64
  barbatum 3(2): 66
    subsp. *lawrenceanum* 3(2): 67
  bougainvilleanum 3(4): 163-164, 164d,
    164pl
  dayanum 3(1): 5d
  druryi 11(1): 35
  exul 10(1): 5-8, 6pl, 8d
  fairrieanum 2(4): 351-354, 352pl, 353d
    var. *album* 2(4): 352-353
    var. *bohlmannianum* 2(4): 352-353
  fowliei 3(1): 66, 67
  glanduliferum 3(4): 159
    var. *glanduliferum* 3(4): 159
    var. *wilhelminae* 3(4): 159, 160
  hennisianum 3(2): Cover pl,
    66-7, 66pl, 67d
  malipoense 4(4): 157-160, 158pl, 159d
  papuanum 3(4): 160
  praestans subsp. *wilhelminae* 3(4):160
  rothschildianum 2(2): 276, 277d
  sanderianum 1(1): 3-6, 4d, 4pl
  schoseri 12(2): 62-66, 64pl, 65d
  sukhakulii 1(2): 81, 83
  urbanianum 3(3): 102-103, 102pl, 103d

  violascens 3(4): 161-162, 162d, 162pl
    var. *gauierense* 3(4): 161-162
  wentworthianum 3(4): 160, 164-166,
    166d, 166pl
  wilhelminae 3(4): 160
  xerophyticum 13(3): 132
  sect. *Paphiopedilum* 10(1): 5
Papilionaceae 14(4): 179
Papilionopsis stylidioides 3(4): 182
Papuacedrus papuana 12(2): 66-72, 68pl,
    69d
    var. *arfakensis* 12(2): 70
    var. *papuana* 12(2): 71
  arfakensis 12(2): 71
  torricellensis 12(2): 71
Papyrus 11(1): 6
Paradoxical Onion 4(4): 194-199
Paramongaia superba 14(3): 142
  weberbaueri 14(4): 142-147, 142pl, 146d
Parasyringa sempervirens 8(2): 59
Parc de Tsimbazaza Botanical Garden
    13(3): 163-164
Parker, D. 1(4): 147
Parker, Patricia 11(3): 149d
Parkinson, S. 14(2): 87-100
Parochetus africanus 8(2): 54-58, 54pl,
    57d
  communis 8(2): 54, 56, 57d
Parthenocissus himalayana 14(3): 157
Paschalococcus dispersa 14(4): 227
Paspalum forsterianum 14(4): 227
Passiflora, National Collection of 15(2):
    114
Passiflora 13(2): 63-69
  amicorum 13(2): 64
  araujoi 13(2): 64
  coccinea 13(2): 63-64, 66
  cuneata 15(2): 112
    'Miguel Molinari' 15(2): 109-114, 110pl,
    113d
  glandulosa 13(2): 63-64, 66
  lourdesae Molinari ined. 15(2): 112
  mixta 11(4): 183

mollissima 11(4)183-186, 194d, 184pl
quadrifaria 13(2): 63-69, 64d, 64pl, 67d
quadriglandulosa 13(2): 64
speciosa 13(2): 64
tholozanii 13(2): 64
tomentosa var. *mollissima* 11(4): 185
tripartita 11(4): 183
tripartita var. *mollisssima* 11(4): 185
vitifolia 13(2): 63-64
sect. *Tacsonia* 11(4): 183, 184
subg. *Distephana* 13(2): 63-64, 66
Passifloraceae 11(4): 183; 13(2): 63
Paton, Alan 8(4): 168-71; 11(3): 109-117
Pediocactus knowltonii 1(4): 182-3, 183d
Pedrola-Monfort, Joan 15(1): 67-77
Pelargonium alchemilloides 10(3): 128
endlicherianum 10(3): 124-129, 126pl, 128d
quercetorum 10(3): 125, 129-131, 130pl, 130d
Peltiera 14(4): 232
Pentapetes erythroxylon 15(1): 29, 33
melanoxylon 15(1): 29
Pentastemona 12(2): 98-99, 98d
Pentastemonaceae 12(2): 97-99
Periboea corymbosa 5(1): 7
gawleri 5(1): 7
Persea americana 1(3): 134
duthiei 1(3): 136-7, 137d
ichangensis 1(3): 136
thunbergii 1(3): 137
Phaenocoma 13(2): 56-62
prolifera 13(2): 56-62, 56pl, 60d
Phaeus tankervillae 11(1): 36
Phalaenopsis amabilis 13(3): 161
Phanera 11(4): 167
Phanerosorus 12(2): 96, 99-100
Phegopteris bukoensis 12(4): 197
levingei 12(4): 197
Philadephus laniger 12(4): 189
Philcox, David 1(1): 2, 6-8; 2(1): 198-201; 2(4): 354-357

Philesia magellanica 4(3): 136d
Philippia 13(3): 214
Philodendron billietiae 13(1): 14-18, 14pl, 17d
Photinia niitakayamensis 5(4): 148-151, 148pl, 149d
undulata var. *formosana* 5(4): 150
Phragmipedium 13(3): 130
besseae 6(4): 156-159
xerophyticum 13(3): 132
Phragmites kharka 13(2): 98
Phu Luang Wildlife Sanctuary 1(2): 75-84
Phylica 13(4): 214
Phyllachne colensoi 7(4): 194d
Phyllitis scolopendrium 7(1): 38d
Phyllocladus hypophyllus 2(2): 282
Phyllostachys 12(1): 8-15
aurea 12(1): 10, 12-13
aureosulcata 12(1): 12, 12pl, 14
f. *alata* 12(1): 14
f. *aureocaulis* 12(1): 14
f. *spectabilis* 12(1): 14
bambusoides 12(1): 10, 12-13, 15
f. *marliacea* 12(1): 9
dulcis 12(1): 10, 14, 14pl
edulis 12(1): 9-10, 12-13
f. *heterocycla* 12(1): 10
elegans 12(1): 9
flexuosa 12(1): 12
heterocycla 12(1): 12
nigra 12(1): 10-13
pubescens 12(1): 9, 12
purpurata 12(1): 9
viridi-glaucescens 12(1): 12, 10pl, 14
viridis 12(1): 9, 12
vivax 12(1): 15
Phylloxylon 14(4): 232, 234, 235, 237
Phymatosorus 12(2): 91
Physurus spp. 13(3): 155
Picea chihuahuana 3(3): 120-122, 123-124
mexicana 3(3): 120d, 122-124
smithiana 13(2): 99d, 101
Pickeringia 14(4): 182

Pigeon pea 14(4): 189
Pilea forgeti 1(1): 17
  microphylla 1(1): 17
  peperomioides 1(1): 14-19, 16d, 16pl;
    1(3): 144; 2(3): 334-336; 14(1): 55-56
Pilgerodendron 12(2): 67
Piliostigma 11(4): 127
Pilosocereus 13(2): 71, 76
  gounelli 9(3): 108d
Pinkroot 9(3): 135
Pinellia tripartita 5(1): 18-21, 18pl, 20d
Pinus wallichiana 13(2): 99, 101; 14(3):
    157
Piptanthus 14(4): 182
Pitcher plants (*Sarracenia*) 15(2): 137
Pithecellobium 14(4): 199
  halogens 14(4): 200
  diopodum 14(4): 200
  pseudo-tamarindus 14(4): 200
Pittosporum gayanum 5(2): 85d
Placseptalia 13(2): 22
Plantago robusta 2(4): 372
Plant availability 2(1): 193
Plant Diversity: Kew's role in north-east
  Brazil 9(3): 103-116
Plant-hunting in the mountains of
  Ecuador 6(4): 172-181
Plant hunting in the Nara Prefecture of
  Honshu, Japan 10(4): 184-190
Plants in Peril 2(4): 380; 3(2): 88-90; 3(3):
    140-2; 4(2): 97-101, 4(4): 199-201;
    5(2): 89-92, 5(3): 137-40; 7(2): 90-5;
    8(2): 82-94, 8(4): 197-202; 9(3): 138-
    42; 10(1): 52-54; 10(3): 144-148; 11(2):
    88-94; 11(3): 146-151, 149d; 12(1): 42-
    45; 13(1): 33-35; 15(1): 62-67; 15(2):
    141-146
Platunium rubrum 13(2): 80
Platylepis glandulosa 8(2): 79d
Platylobium 11(4): 172
Platymiscium 14(4): 187
Plectocephalus 13(1): 3-4
  americana 13(1): 3

chilensis 13(1): 3-4
rothrockii 13(1): 3-7, 4pl, 6d
varians 13(1): 4
Plectrelminthus spiculatus 4(2): 73
Pleione aurita 5(4): 162, 163-165, 164pl,
    164d
  limprichtii 3(2): 73-78, 74d
  scopulorum 5(4): 165-167, 166pl, 166d
  cultivars 1(1): 48
Pluchea 13(4): 196
Poitea 14(4): 182
Polhill, Roger M. 8(2): 54-58; 14(4): 176-
    183, 208-212
Pollination, bird 11(4): 179-206
Polypodiaceae 12(2): 89-95
Polypodium cordifolium 8(3): 115
  lomarioides 12(2): 92-93
  sarcopum 12(2): 93
Polystachya sect. *Cultriformes* 5(1): 22
  melliodora 5(1): 22-4, 22pl, 23d
Polyxena corymbosa 5(1): 5-7, 6pl, 6d
Pongamia 14(4): 176
Ponsonby, Laura 7(3): 138-142
Populus ciliata 13(2): 101
Porter, Colin 6(4): 172-81
Portunus pelagicus 15(4): 268d
Pothos miniata 12(2): 87
Poupartia castanea 13(4): 232d, 233
Prangos eryngioides 7(3): 124
Pratia physaloides 7(4): 197d
Prendergast, Hew D.V. 9(2): 76-85
Price, Valerie 2(4): 346pl; 3(1):
  Cover pl, 14pl, 20pl, 22pl, 28pl,
    32pl, 34pl; 3(2): 52pl; 4(2): 62pl,
    70pl, 74pl, 108pl; 5(1): 6pl; 6(1):
    20pl, (caption in error-see p.144);
    6(2): 62pl; 7(3): 4pl, 7(4): 176pl; 13(4):
    196d, 198d; 15(2): 92pl, 94d
Pridgeon, Alec M. 10(1): 46-51
Primula aureata 5(1): 31d
  denticulata 13(2): 98
  longipes 2(4): 364, 364d
  marginata 4(1): 55d

41

wollastonii 1(1): 33, 33d; 14(3): 158, 159d
Primulaceae 14(3): 137; 15(3): 180
Prosopis 14(4): 176
Protea aurea subsp. *aurea* 11(4): 174-178,
   174pl, 177d
  calycina 11(4): 176
  latifolia 11(4): 175, 176
  longiflora 11(4): 176
  neriifolia 11(4): 174
  ovata 11(4): 176
  radiata 11(4): 176
  repens 11(4): 174
Proteaceae 11(4): 174, 178
Proteas 3(1): 45
Prunus cornuta 13(2): 101
Pseudochadsia 14(4): 234
Pseudogynoxys benthamii 9(4): 155
  cabrerae 9(4): 153-156, 154pl, 154d
Pseudophegopteris 12(4): 195
  levingei 12(4): 195-197, 196pl
Psiguria warscewiczii 11(3): 151
Pterocarpus 14(4): 182
Pterostyrax corymbosa 10(4): 189d
Ptychosema 11(4): 172
Purves, Rodella 1(1): 2, 4pl, 10pl; 5(3):
  102pl, 104pl, 110pl,112pl, 114, 116pl
Puschkinia ? *dubia* 11(3): 130
Pycnospatha arietina 10(3): 121-124,
  122pl
  soerensenii 10(3): 123
Pygmy sundews 10(3): 139
Pyranthus 14(4): 232
Pyrolirion aurantiacum 6(3): 101
  aureum 6(3): 101
  flammeum 6(3): 101
  tubiflorum 6(3): 95-101, 96pl, 100d

Quercus brantii 2(4): 362
  in Turkey 2(4): 362m
  libani 2(4): 362
  semecarpifolia 13(2): 101
Quesnelia humilis 8(2): 69-71, 70pl, 70d
  tillandsioides 3(4): 182, 183

Radcliffe-Smith, Alan 1(1): 2, 14-19; 4(3):
  110-113; 14(1): 55-56
Rafflesia tuan-mudae 2(2): 275, 275d
Raimondo, F.M. 13(2): 105-107
Raistrick, Reinhild 15(1): 52pl
Ramosmania 13(4): 204-209, 204pl, 208d
  heterophylla 13(4): 204-207
  rodriguesii 13(4): 204-209
Randia 13(4): 204
Randjelovic, Novica 7(4): 182-186
Randjelovic, Vladimir 7(4): 182-186
Randrianasolo, Voara 7(2): 90-95
Ranunculaceae 15(4): 231, 242
Ranunculus weberbaueri 15(4): 231-235,
  232pl
Rapa Nui (Easter Island) 14(4): 221, 226-
  231
Rapuntium blandum 10(2): 73
  bridgesii 10(2): 73
  lucaeanum 10(2): 73
  tupa 5(4): 161
Rara National Park, Nepal 13(2): 96-104
Ravenala madagascariensis 13(4): 222
Ray, Susie 2(1): 202pl, 204pl
Read, Mike 6(1): 22-29
Recently introduced species of
  Pleione from China 5(4): 161-167
Recovery from the 1987 storm at
  Wakehurst Place 6(3): 134-139
Redwood 15(1): 29
Rees-Davies, Kay 14(4): 222pl
Reeve, Tom 2(3): 291-308
Reginald Farrer and Glasnevin
  10(4): 171-178
Reid, Clare 11(1): 9-13
Réunion 13(4): 182-185, 215-216, 217-233
Renvoize, Stephen 8(4): 185-194; 12(1): 8-
  15
Reuter, G.-F. 11(3): 134-135, 137d
Raphidophora 12(2): 87
Rhanterium 13(4): 195
Rhaphidorhynchus batesii 11(2): 58

spiculatus 4(2): 69, 73
Rheopteris cheesmanii 12(2): 100, 101d
Rhipsalis baccifera 14(3): 125
  brachiata 14(3): 126
  cereuscula 14(3): 126
  pilocarpa 14(3): 125-129, 126pl, 128d
  subg. *Erythrorhipsalis* 14(3): 126
Rhizanthella gardneri 1(2): 84-91, 87d;
  10(4): 163
  slateri 1(2): 84-91
Rhododendron acrocline 12(2): 82
  angiense 12(2): 82
  anthopogon 13(2): 103
  arboreum 13(2): 101; 14(3): 157
  brookeanum 9(1): 18-21, 18pl
    var. *flavum* 9(1): 19
  burttii 6(3): 110-112, 110pl, 111d
  campanulatum subsp.*aeruginosum*
    7(1): 10
  cantabile 7(4): 178
  christi 2(1): 208-210, 208pl, 210d
    var. *loniceroides* 2(1): 210
  coenenii 12(2): 82
  cowanianum 5(1): 28d, 29
  cuffeanum 12(4): 243
  culminicolum 12(2): 77-85
    var. *angiense* 12(2): 77-85, 78pl, 80d,
      83d
    var. *culminicolum* 12(2): 81-82, 83d
    var. *nubicola* 12(2): 80d, 82
  dalhousiae var. *dalhousiae* 4(2): 59-61,
    60pl, 60d
  dielsianum 1(3): 108-111, 108pl, 110d
  discolor 3(2): 51-52, 55
  ericoides 2(2): 283
  fortunei 3(2): 51-52
    subsp. *discolor* 3(2): 51-55, 52pl,54d
  gibbsiae 12(2): 82
  hatamense 12(2): 82
  javanicum var. *teysmannii* 9(1): 19-21
  laureola 1(3): 111
  lepidotum 13(2): 103
  lowii 2(2): 278

  lyi 1(2): 77
  konori 5(3): 113-116, 114pl, 115d
  lineare 5(3): 102-104, 102pl, 103d
  nubicola 12(2): 82
  osmerum 7(4): 178
  pleianthum 4(4): 165-168, 166pl, 167d
  praetervisum 5(3): 104-107, 104pl, 106d
  'Princess Alice' 9(1): 15-18, 16pl
  rarilepidotum 1(1): 20-22, 20d, 20pl;
  russatum 7(4): 176-179, 176pl, 177d
  salicifolium 5(3): 110-113, 112pl, 112d
  stenophylla 2(2): 278
  stenophyllum 5(3): 107-110, 108d,
    110pl
    ssp. *angustifolium* 5(3): 110
    ssp. *stenophyllum* 5(3): 109
    var. *angustifolium* 5(3): 110
  superbum 5(3): 116-119, 116pl, 118d
  teysmannii enigma, The Rhododendron
    9(1): 18-21
  Vireya rhododendrons, natural hybrids
    12(2): 84
  Vireya rhodoendrons, some attractive
    5(3): 99-119
  sect. *Vireya* 12(2): 77-78
  sect. *Javanicae* 12(2): 78
Rhodostachys chamissonis 13(1): 22-23
  elegans 13(1): 25
Rhopalopodium weberbaueri 15(4): 234
Rhus batophylla 10(2): 64-66, 64pl
  javanica 10(4): 187
  subgen. *Rhus* 10(2): 64
  subgen. *Thezera* 10(2): 64
Rhynchospora ciliata 10(3): 120
  exigua 10(3): 120
  jelskiana 10(3): 120
  nervosa 10(3): 117-121, 118pl,119d
    subsp. *ciliata* 10(3): 120
    subsp. *nervosa* 10(3): 120
    var. *ciliata* 10(3): 120
    var. *tenuifolia* 10(3): 120
  persooniana 10(3): 120
  sect. *Dichromena* 10(3): 117

Ribes speciosum 11(4): 204, 204d

Rico-Arce, M. de L. 10(2): 66-69; 11(4): 167-171, 171-174; 14(4): 198-202

Riverine forests of the Jubba river, southern Somalia 5(3): 120-129

Robbins, Sarah 6(4): 156-159; 7(1): 6-9; 8(1): 5-8; 9(3): 116-120; 10(1): 5-8,18-22, 52-54

Robbrecht, Elmer 10(2): 76-80; 15(1): 42-48

Robertson, Ann 6(4): 166-171; 15(1): 49-62

Robinia 14(4): 176, 182

 pseudoacacia 14(4): 192

Robson, N.K.B. 15(2): 92-95

Rolfe, Robert Allen 10(1): 36, 46-51

Rodrigues 13(4): 182-185, 216-233

Rodrigues, J. Barbosa 13(3): 153d, 154d

 Orchid paintings of 13(3): 152-157

Rodriguezia eleutherosepala 13(3): 155

Romulea hirta 1(3): 106-107, 106pl, 107d

Rosa 14(1): 13-15, 64-76

 afzeliana 14(2): 66

 banksiae 14(1): 13, 14

 bracteata 4(1): 49d

 canina 14(2): 66, 70-72

 collettii 14(1): 13

 complicata 14(2): 64-76

 'Complicata' 14(1): 64-76, 64pl

 cymosa 14(1): 13-15

  'Rebecca Rushforth' 14(1): 13-15, 14pl

 dumalis 14(2): 66

 froebelii 14(2): 68

 fugax 14(2): 74

 gallica 14(2): 68-72

 gallica x *dumalis* 14(2): 73

 gallica x *glauca* 14(2): 68-69

 gallica x *glauca* var. *complicata* 14(2): 69

 glauca 6(1): 3-6, 4d, 4pl; 14(2): 66-68

 glauca var. *complicata* 14(2): 69, 73

 glaucescens 6(1): 6

 gutensteinensis 6(1): 6

hugonis 5(1): 39-43

 and Father Hugh 5(1): 39-43

ilseana 6(1): 6

indica 14(1): 13

intricata 14(2): 74

laxa 14(2): 68

'Laxa' 14(2): 68

macrantha 14(2): 72

'Macrantha' 14(2): 72

microcarpa 14(1): 13, 14

mollis 6(1): 3

reuteri 14(2): 66-67, 73-74

 var. *adenophora* 14(2): 74

 var. *genuina* 14(2): 74

 var. *intermedia* 14(2): 73, 74

 var. *transiens* 14(2): 74

rubicunda 6(1): 6

rubrifolia 6(1):6; 14(2): 66

stylosa 6(1):4

sect. *Banksianae* 14(1): 13, 14

sect. *Caninae* 14(2): 72

Rosaceae 12(4): 198-201; 14(1): 13; 14(2): 64; 15(2): 115

Roscoea alpina 11(1): 13, 14, 104; 13(1): 8-9; 14(2): 77-81, 78pl, 80d

 alpina var. *minor* 14(2): 79

 'alpina' 15(4): 227

 auriculata 11(1): 13; 11(3): 105; 13(1): 9; 14(2): 79; 15(4): 222

 australis 13(1): 9

 brandisii 11(3): 105

 capitata 11(3): 104; 13(1): 8-9; 15(4): 226-230, 226pl, 229d

 var. *purpurata* 15(4): 227

 var. *scillifolia* 15(4): 227

 cautleoides 11(1): 13-15; 11(3): 106; 14(1): 3; 14(2): 79; 15(4): 227

 cautleoides var. *schneideriana* 11(1): 14, 16

 debilis 14(1): 3

 debilis var. *limprichtii* 13(1): 9

 exilis 14(2): 79

 ganeshensis 13(1): 8-13, 8pl, 12d

humeana 11(1): 13
intermedia 14(2): 79
intermedia var. *anomala* 14(1): 4
intermedia var. *macrorhiza* 14(1): 4
longifolia 14(2): 79
praecox 14(1): 2-6, 2pl, 5d
procera 11(3): 107
purpurea 11(1): 13; 11(3): 104-107;
　　13(1): 8-10; 14(2): 79; 15(4): 222
　var. *exilis* 14(2): 79
　var. *gigantea* 11(3): 107
　var. *procera* 11(3): 107
　var. *purpurea* 11(3): 107
　var. *sikkimensis* 11(3): 105
purpurea 'Red Gurkha' 11(3): 104-109,
　104pl, 107d; 13(1): 9
schneideriana 11(1): 13-18, 17d, 14pl;
　14(1): 3
scillifolia 11(1): 13; 14(2): 78-79; 15(4):
　227
tibetica 14(2): 78
tumjensis 13(1): 8-9; 15(4): 220-225,
　220pl, 224d
wardii 13(1): 9
yunnanensis 11(1): 14
　var. *dielsiana* 11(1): 14, 16
　x *beesiana* 11(1): 14
Ross-Craig, Stella 12(4): 241
Rossi, Walter 10(1): 43-46; 11(1): 28-31
Rosularia muratdaghensis 9(2): 59-62,
　60pl, 62d
Rothwell, Nigel 8(3): 118-123
Round Island (Mascarene Is.) 13(4): 184,
　186
Rourke, John 11(4): 178-183
Royal Chitwan National Park, The 9(4):
　183-193
Rubiaceae 13(4): 204-209; 15(1): 42
　tribe *Gardenieae* 13(4): 204
Rubus, National Collection 15(2): 119
Rubus alceifolius 13(4): 225
　coronarius 15(2): 118
　pedatus 2(1): 226

procerus 'Himalayan Giant' 5(1): 32-36
rosaefolius var. *coronarius* 15(2): 118
rosifolius 15(2): 118
　'Coronarius' 15(2): 115-119, 116pl
　var. *commersonii* 15(2): 116
saxatilis 2(3): 316d
　sect. *Idaeobatus* 15(2): 117
　subsect. *Rosaefolii* 15(2): 117
Ruckia 13(1): 22
Rudbeckia hispida 4(1): 34
　Purpurea 4(1): facsimile t.2, 34
　Serotina 4(1): 34
　Speciosa 4(1): 34
Rushforth, Keith 3(3): 119-24
Russell, T.A. 10(3): 133
Rutaceae 12(1): 25-28; 15(2): 175
　subtribe *Toddaliinae* 4(4): 173

Sabinea 14(4): 182
Saccolabium mooreanum 12(4): 239
Saffron (*Crocus*): 2(3): 309
Sagamartha National Park, Nepal 14(3):
　153-166
Sailor caps 14(3): 137
Sainsbury Orchid Conservation Project
　10(1): 38-43
St Helena 2(4): 370m
　endemic plants 2(4): 369-379
Saintpaulia, conservation status of 15(1):
　49-62
Saintpaulia 1(2): 92-93; 15(1): 49-62
　amaniensis 15(1): 52
　confusa 15(1): 51
　difficilis 15(1): 54
　grotei 15(1): 51
　ionantha 1(2): 92; 15(1): 50d, 51
　magungensis 15(1): 52pl
　rupicola 15(1): 56
　tongwensis 15(1): 54, 64pl, 65
　cultivars 15(1): 51
　sp. nov. 'Kacharoroni' 15(1): 56
Salicaceae 12(4): 207-209
Salisbury, Sir Edward James 10(3):133

Salix 12(4): 207
 helix 12(4): 207
 koriyanagi 12(4): 207-209, 208d, 208pl
 purpurea 12(4): 207-208
  var. *japonica* 12(4): 208
Salvia 8(4): 168
 cardinalis 11(2): 52
 castanea 1(2): 96
 darcyi 11(2): 52-55, 52pl, 54d
 diversifolia 8(4): 170
 fulgens 11(2): 52
 lanceolata 8(4): 168-171, 168pl, 170d
 nivea 8(4): 170
 nitida 8(4): 170
 stolonifera 11(2): 52
 sect. *Cardinales* 11(2): 52
 sect. *Fulgentes* 11(2): 52
 sect. *Holwaya* 11(2): 52
 subg. *Calosphace* 11(2): 52
Samania 14(4): 199
Sampson, Hugh Charles 10(3): 132
Sander, Frederick 13(3): 134
Sands, Martin 3(3): 112-116; 7(2): 57-85
São Raimundo Nonato, Fundação Ruralista
 2(1): 215m
Sarcopodium 13(3): 139-140
Sarracenia, National Collection of 15(2):
 137
Sarracenia alata 15(2): 138
 flava 15(2): 138
  var. *ornata* 15(2): 138
  var. *rugelii* 15(2): 138
  cultivars 15(2): 138
 flava var. *catesbaei* 10(3): 144
 flava var. *oreophila* 10(3): 144
 leucophylla 15(2): 139
  'White Trumpet' 15(2): 138pl, 139
 minor 15(2): 138pl, 139
  'Okefenokee Giant' 15(2): 139
 oreophila 10(3): 144-148, 146d; 15(2):
  138pl, 139
 psittacina 5(2): 60-65, 60pl, 64d; 15(2):
  139

 purpurea 15(2): 139
  subsp. *purpurea* 15(2): 139
  subsp. *venosa* 15(2): 139
  'red forms' 15(2): 138
 rubra 15(2): 139
  subsp. *alabamensis* 15(2): 140
  subsp. *gulfensis* 15(2): 140
  subsp. *jonesii* 15(2): 138pl, 140
  subsp. *rubra* 15(2): 140
  subsp. *wherryi* 15(2): 140
Sassafras albidum 1(3): 133
Satyrium hircinum 11(1): 26
Saxegothaea conspicua 4(3): 131
Saxifraga florulenta 2(1): 232-234, 233d
 lychnitis 1(1): 34d, 35
 rosacea 2(3): 315d
Scheer, Frederick 11(2): 70, 74-81, 75p
Scheeria 11(2): 80
Schickendandtzia pymaea 14(1): 7
Schilling, Tony 1(3): 99, 133-139; 5(1):
 24-32; 6(2): 65-73; 9(4): 183-193;
 13(2): 96-104; 14(3): 153-166
Schima wallichii 2(2): 279
Schlumbergera 14(3): 125
Schmidt, F. 14(1): 40
Schoenus ciliatus 10(3): 120
 tenuifolius 10(3): 120
Schott, H.W. 12(3): 154-157, 155p
Schrenk, L. von 14(1): 40
Schrire, B. 14(4): 184-193, 213-220
Schulze, M. 14(2): 68-69
Schultesia crenuliflora 9(3): 111d
Scilla bifolia 4(4): 195
 brevifolia 5(1): 7
 corymbosa 5(1): 7
 maderensis 5(1): 11-15, 12pl, 14d
  var. *melliodora* 5(1): 12
 paradoxa 4(4): 195, 196
Sclerocactus scheeri 11(2): 80
Scorpiris 3(2): 61
Scrophulariaceae 12(4): 202-206
Scyphochlamys 13(4): 204-205
 revoluta 6(3): 102-106, 102pl, 103d;

13(4): 204
Scyphostegia 12(2): 98-99, 98d
Scyphostegiaceae 12(2): 97-99
Seaforthia elegans 11(2): 77
Sedum lucidum 4(2): 64-66, 64pl, 66d
Seemannia latifolia 12(4): 192
  sylvatica 12(4): 192
Sellars, Pandora 1(2): 50, 56pl, 60pl, 64pl,
    66pl; 2(1): 196pl, 212pl; 2(3): 296pl,
    298pl, 300pl, 302pl, 304pl, 306pl; 2(4):
    348pl, 352pl; 3(2): 56pl, 66pl; 3(3):
    102pl; 3(4): 162pl, 164pl, 166pl; 4(2):
    68pl; 5(1): 22pl, 5(2): 72pl; 7(1): 6pl;
    10(1): 6pl, 10pl; 13(4): 200pl; 14(3):
    148pl; 15(2): 116pl
Senecio benthamii 9(4): 155
  cervilingua 8(1): 3
  erthractis 8(1): 3
  formosus 8(1): 1-5, 2pl, 4d
    var. *angustifolius* 8(1): 3
    var. *latifolius* 8(1): 3
    var. *subruncinatus* 8(1): 3
  leucadendron (He-Cabbage): 2(4): 377
  redivivus 2(4): 376-377
  tabacon 8(1): 3
  rutshuruensis 7(3): 112
  seretii 7(3): 112
  subalatipetiolatus 7(3): 112
  variostipellatus 7(3): 112
Senna 14(4): 176, 177, 181
  candolleana 5(2): 79, 80d
  corymbosa 5(2): 77, 78d, 80d
  x *floribunda* 5(2): 77d, 80, 80d
  multiglandulosa 5(2): 80
  septentrionalis 5(2): 79, 81
  Leaf shapes in 5(2): 80d
SEPASAL 10(3): 135
SEPASAT 10(3): 134
Serapias 10(4): 163
  algarbiensis 3(1): 25
  cordigera subsp. *Orientalis* 3(1): 28
  excavata 3(1): 25
  glabra 3(1): 25

lingua 3(1): 22-25, 22pl, 24d
  subsp. *durieui* 3(1): 25
  subsp. *excavata* 3(1): 25
  subsp. *oxyglottis* 3(1): 25
  subsp. *stenopetala* 3(1): 25
    var. *durieui* 3(1): 25
    var. *stenopetala* 3(1): 25
longipetala 3(1): 28
mauretanica 3(1): 25
orientalis subsp. *apulica* 3(1): 28
  subsp. *orientalis* 3(1): 28
    var. *cordigeroides* 3(1): 28
oxyglottis 3(1): 25
pseudocordigera 3(1): 28
stenopetala 3(1): 25
vomeracea f. *platyglottis* 3(1): 28
  f. *platypetala* 3(1): 28
  subsp. *orientalis* 3(1): 22pl, 25-28, 26d
Serapiastrum lingua 3(1): 25
Serianthes 14(4): 199
Sesbania 14(4): 182
Shackelton, Lydia 12(4): 239; 15(4): 280
Shaffer-Fehre, Monica 12(2): 101d
Shamrock Pea 8(2): 54
She-Cabbage (*Lachanodes arborea*):
  2(4): 376-377
Sherlock, Siriol 10(3): 112pl; 14(1): 14
Shine, Charles 15(3): 158-163
Shooting star 14(3): 137
Shorea robusta 9(4): 189
Shortia soldanelloides 2(1): 226
Siberia 14(1): 35-55
Sideroxylon 13(4): 185, 214
  spinosum 9(2): 76
Signorini, Maria Adele 11(1): 28-31
Simmons, John 3(1): 39-48
Simpson, David 6(1): 29-37; 10(3):
  117-121; 11(1): 6-9
Simpson, Nikki 15(3): 186pl, 191d, 192d
Simpson, Rosemary B. 8(2): 89-94;10(3):
  144-148; 12(2): 111-119
Sinningia tubiflora 12(4): 239
Siphocampylus andropogon 9(1): 5

47

macranthus 9(1): 5
spectabilis 9(1): 5
surinamensis 9(1): 5
Sir Joseph Banks Building 10(3): 135
Sir John Kirk's Field Drawings at Kew
    4(3): 141-144
Sir William's (Hooker) Irish Lace 12(4):
    220-236
Sisyrinchium, National Collection of 15(2):
    108
Sisyrinchium macrocephalum 15(2): 106
    subsp. *fuscoviride* 15(2): 104
    subsp. *giganteum* 15(2): 105
    palmifolium 15(2): 104-108, 104pl, 107d
Skimmia 12(1): 25-28, 31-33
    A Revision of the Genus 4(4): 168-194
    anquetilia 4(4): 190-191; 13(2): 101
    arborescens 4(4): 184-187
        subsp. *arborescens* 4(4): 185
        subsp. *nitida* 4(4): 185
    arisanensis 4(4): 183
    x *confusa* 4(4): 191-193
    'Chelsea Physic' 6(4): 159-162,
        160pl, 161d
    'Isabella' 4(4): 192
    'Kew Green' 4(4): 172
    distincte-venulosa 4(4): 183
    euphlebia 4(4): 184
    fortunei 4(4): 182
        var. *argentea* 4(4): 183
    fragrans 4(4): 182
    fragrantissima 4(4): 182
    hainanensis 4(4): 183
    intermedia 4(4): 182
    japonica sensu Lindley 4(4): 183
    japonica Thunb. 4(4): 178-184
        subsp. *Japonica* 4(4): 181
        subsp. japonica 'Wakehurst White'
            12(1): 25-28, 26pl, 27d
        subsp. *lutchuensis* 4(4): 182
        subsp. *reevesiana* 4(4): 182
        subsp. *reevesiana* 'Robert Fortune'
            12(1): 32d

    var. *argentea* 4(4): 183
    var. *argentea-variegata* 4(4): 183
    var. *euphlebia* 4(4): 184
    var. *intermedia* 4(4): 181
    var. *japonica* 4(4): 182
    var. *kwangsiensis* 4(4): 183
    var. *reevesiana* 4(4): 183
    'Chilan Choice' 4(4): 180, 180pl
    'Robert Fortune' 4(4): 180
    var. *repens* 4(4): 181
    var. *rugosa* 4(4): 182
    var. *variegata* 4(4): 183
    f. *repens* 4(4): 181
    f. *rugosa* 4(4): 182
    kwangsiensis 4(4): 183
    laureola (DC.) Sieb. & Zucc. ex Walp.
        4(4): 187-190
        subsp. *lancasteri* 6(4): 162
        subsp. *laureola* 4(4): 189; 6(4): 162
        subsp. *multinervia* 4(4): 188pl, 189
    laureola auctt. 4(4): 191
    leaf outlines 4(4): 177d
    lutchuensis 4(4): 182
    melanocarpa 4(4): 185, 189; 6(4):162
    multinervia C.C.Huang 4(4): 189
    multinervia auctt. 4(4): 193
    oblata 4(4): 182
    orthoclada 4(4): 183
    philippinensis 4(4): 183
    reevesiana 4(4): 182
        var. *variegata* 4(4): 183
    repens 4(4): 181
        var. *rugosa* 4(4): 182
    rubella 4(4): 182
    rugosa 4(4): 182
    veitchii 4(4): 182
    wallichii 4(4): 185, 189
Skottsbergiella 3(4): 154
    Fernandezia 3(4): 156
Small, David 2(3): 324-330
Smilax megalantha 3(3): 116-118, 116pl,
    118d
Smith, A. 10(2): 87

48

Smith, Gideon F. 14(1): 16-22
Smith, Louise 14(3): 138pl
Smith, Sarah 15(3): 158-163
Snow, D.W. 11(4): 198-206
Solander, David 14(2): 87-88
Solanopteris 12(2): 90
Solenotinus foetens 14(3): 122
Sonder, Otto Wilhelm 13(1): 36
Sophora 14(4): 180, 182, 221-226
  denudata 13(4): 216
  japonica 14(4): 221-222, 242
  macrocarpa 14(4): 222
  microphylla 14(4): 222
  prostrata 14(4): 222
  tetraptera 14(4): 222, 224
  toromiro 14(4): 192, 221-231, 222pl,
    225d
  sect. *Edwardsia* 14(4): 222
Sparrow, Mark 14(4): 213-220
Spathoglottis plicata 6(3): 128d
Spigelia 9(3): 134
  flava 9(3): 134-138, 134pl, 137d
  marilandica 9(3): 134
Sophronitis 2(3): 296
Spice Bush (*Lindera benzoin*): 1(3): 136
Spondias tuberosa 2(1): 223
Sprekelia formosissima 14(2): 93d
Spruce (*Picea*):
  Chihuahua (*P. chihuahuana*): 3(3): 120-
    122, 123-124
  Mexican (*P. mexicana*): 3(3): 120d,
    122-124
  of Mexico 3(3): 119-124
Sprunger, Samuel 13(3): 152-157
St Helena 15(1): 27-28
  endemic flora and fauna of 15(1): 28
  ebony 15(1): 27, 29
Stachytarpheta 13(1): 33-35
  candida 13(1): 33
  cayennensis 13(1): 33
  chamissonis 13(1): 33
  crassifolia 13(1): 33
  gesnerioides 13(1): 33

  jamaicensis 13(1): 33
  macedoi 13(1): 33
  sellowiana 13(1): 33-35, 34d
  sericea 13(1): 33
Stamenkovic, Vlastimir 7(4): 182-186
Staniforth, Martin 13(2): 58-59; 13(4):
  192, 197, 210, 206-207, 212-213; 14(4):
  213-220, 221-226; 15(1): 27-36; 15(3):
  186-193; 15(4): 236-241
Stannard, Brian 5(4): 151-153; 9(3): 128-
  131
Staudtia gabonensis 13(3): 168
Stearn, William T. 4(4): 194-199; 9(4):
  180-182; 10(4): 178-184; 12(1): 15-25
Stenoglottis, The genus 6(1): 9-22
Stenoglottis fimbriata Lindl. 6(1):
  12-15, 12pl, 13d
  fimbriata Summerhayes 6(1): 20
  longifolia 6(1): 20-2, 20pl, 20d
  woodii 6(1): 15-17, 16pl, 16d
  zambesiaca 6(1): 17-20, 18pl, 19d
Stephanocereus 13(2): 71-72
  leucostele 13(2): 71
  luetzelburgii 13(2): 71
Sterculia 11(2): 88
Sterculiaceae 11(2): 88-89; 15(1): 27
Stereosandra javanica 6(3): 132
Steudnera 12(3): 145
  colocasiifolia 12(3): 145
    var. *discolor* 12(3): 146
  discolor 12(3): 145-147, 146d, 146pl
Stewart, Joyce 4(2): 69-74; 5(1):
  15-18, 5(2): 51-55; 6(1): 9-22; 10(1):
  38-43
Stoebe 13(2): 57
Stone, Judi 13(3): 146pl, 151d; 15(1):
  20pl, 25d
Stones, Margaret 1(1): 12pl; 1(4): 146;
  15(4): 236
Strahm, Wendy 13(4): 182-185, 214-237
Strangweia 11(3): 128
  spicata 11(3): 130
Strangweja 11(3): 128

49

hyacinthoides 11(3): 128-129
  spicata 11(3): 128
Stranvaesia 11(3): 128
Stranvaesia davidiana 2(2): 279
  nussia 2(2): 279
  niitakayamensis 5(4): 150
Streblorrhiza speciosa 14(4): 230; 15(2):
  127-128, 128pl(b/w)
Streptocarpus in Malawi 11(2): 82-88
Streptocarpus 11(2): 88
  arcuatus 11(2): 82, 86
  buchananii 11(2): 83
  dolichanthus 11(2): 83, 86
  erubescens 11(2): 82, 85
  eylesii 11(2): 82
    subsp. *brevistylus* 11(2): 82, 86-87
    subsp. *eylesii* 11(2): 82, 87
    subsp. *silvicola* 11(2): 82, 86
  goetzei 11(2): 83, 85, 87
  hirsutissimus 6(2): 83-84
  hirtinervis 11(2): 83, 85
  leptopus 11(2): 83, 86
  michelmorei 11(2): 88
  milanjianus 11(2): 83, 85-86
  nimbicola 11(2): 83, 85
  solenanthus 11(2): 83, 87
  wittei 11(2): 83, 87
  sp. aff. *michelmorei* 11(2): 83, 88
  subg. *Streptocarpella* 11(2): 82-83
  subg. *Streptocarpus* 11(2): 82-83
Strombocactus disciformis 1(4): 184-185
Strongylodon 1(4): 188-190; 14(4): 176,
  182, 234
  macrobotrys 1(4): 188-190, 189d
Stryphnolobium japonicum 14(4): 221
Stuart-Smith, Susanna 10(3): 146d
Summerhayes, Victor S. 10(1): 36
Surenus glabra 15(3): 189
  serrata 15(3): 189
  serrulata 15(3): 189
  sinensis 15(3): 189
Sutherlandia 14(4): 182
Swainsonia 14(4): 182

Sweet Olives of Saudi Arabia 5(1): 36-38
Synnott, Donal M. 12(4): 195-197
Syringa 8(2): 58-61
  afghanica 6(3): 117
  laciniata 6(3): 116-118, 120-123
  x *laciniata* 6(3): 120, 121, 122pl
  oblata 6(3): 123
  persica 6(3): 116, 122
  x *persica* 6(3): 122
  protolaciniata 6(3): 121-124
  'Kabul' 6(3): 120, 122pl
Syzygium 13(4): 185

Tacsonia mollissima 11(4): 185, 194d, 195
Taeniophyllum coxii 8(2): 79d
Taiwan Beech 12(1): 42-45
Tamarindus 14(4): 181
  indica 14(4): 235
Tanzania: Eastern Arc forests 3(2): 83-87
Tarout (*Cupressus dupreziana*): 3(2): 88-
  90, 89d
Tartu, Estonia 14(1): 35-55
Tassili Cypress (*Cupressus dupreziana*):
  3(2): 88-90, 89d
Taxus baccata subsp. *wallichiana* 13(2):
  101
Taylor, Nigel 1(4): 146, 147-180; 2(2):
  243-273; 3(4): 156-158; 4(4): 168-194;
  5(4): 148-151; 6(1): 3-6; 6(2): 55-58;
  7(4): 173-176; 9(3): 138-142; 11(2): 70-
  74; 12(1): 25-28; 13(2): 70-78; 14(1):
  13-15; 14(3): 120-124, 125-129
Taylor, Peter 10(1): 36
Tea Tree (*Dendrobium canaliculatum*):
  1(3): 128
Tebbs, Margaret 12(2): 83d; 13(4): 193d;
  14(4): 218d, 233d, 236d, 238d; 1515(2):
  107d, 167d
Tecophilaea cyanocrocus 1(1): 24, 24d;
  2(2): 285-288, 286d; 15(2): 126-127
  var. *leichtlinii* 2(2): 286
  var. *violacea* 2(2): 286
Tectaria prolifera 8(3): 146

Tectiphiala 13(4): 186-187
Tephrosia 14(4): 181, 237
  phylloxylon 14(4): 237
Tetradium 5(4): 181-191
  daniellii 5(4): 182, 183, 184d, 189
  fraxinifolium 5(4): 184d, 188
  glabrifolium 5(4): 184d, 185
  ruticarpum 5(4): 186d, 187
Tetraplacus taubertii 8(3): 165
Thailand, Phu Luang Wildlife Sanctuary
  1(2): 76m
Thamnocalamus spathaceus and its
  hundred year flowering cycle 8(4): 185-
  194
Thamnocalamus aristatus 8(4): 192
  cuspidatus 8(4): 193
  falconeri 8(4): 192
  spathaceus 8(4): 185-194, 186pl
  tessellatus 8(4): 193
Theilade, Ida 12(2): 73-77
Thelypteridaceae 12(4): 195-197
Thelypteris levingei 12(4): 197
Thelysia alata 3(2): 65
  grandiflora 3(2): 65
  tarhunensis 3(2): 65
Thermopsis 14(4): 181
Thespesia danis 5(3): 126d
Thiselton Dyer, Lady Harriet 13(3): 155,
  156d
Thomas Moore, The fern herbarium of
  8(3): 147-154
Thomas, Sarah 10(4): 164d; 13(2): 82-85;
  13(3): 134-138, 137d; 13(4): 202d
Threatened Plants Committee 1(4): 182-
  184
Thuja 12(2): 66
  papuana 12(2): 70
Thymelaeaceae 11(3): 117
Thymus cephalotus 11(3): 109
  longiflorus 11(3): 109
Tillandsia 6(1): 26, 28
  caput-medusae 6(1): 23d
  filifolia 6(1): 27d

  ionantha 6(1): 25d
  pruinosa 6(1): 27d
Tipuana 14(4): 176, 182
Tmesipteris 5(3): 133-135
  tannensis 5(3): 133
Tontanea guianensis 10(2): 79
Toona glabra 15(3): 189
  longifolia 15(3): 189
  microcarpa var. *denticulata* 15(3): 190
    var. *grandifolia* 15(3): 190
  serrata 15(3): 189
  serrulata 15(3): 189
  sinensis 15(3): 186-193, 186pl, 191d,
    192d
   'Flamingo' 15(3): 186
  var. *denticulata* 15(3): 190
  var. *glabra* 15(3): 189
  var. *grandifolia* 15(3): 190
  var. *grandis* 15(3): 190
  var. *hupehana* 15(3): 190
  var. *incarvillei* 15(3): 190
  var. *puberula* 15(3): 190
  var. *schensiana* 15(3): 190
Toromiro Management Group 14(4): 226-
  230
Toromiro tree 14(4): 221, 231
Toscano de Brito, A.L. 9(3): 116-120;
  13(3): 152-157
Trautvetter, E.C. von 14(1): 40
Tree fern (*Dicksonia*): 2(4): 378
Tree-ferns 5(3): 129-135
Trevithick, William E. 12(4): 186pl, 187;
  14(3): 168d
Trichomanes lorencei 13(4): 233
Trichosporum angustifolium 8(1): 22
  tetraquetrum 8(1): 22
Tridactyle 15(4): 249
  truncatiloba 13(3): 170, 171d; 15(4):
    249-254, 250pl, 253d
  subg. *Eu-tridactyle* 15(4): 249
   sect. *Eu-tridactyle* 15(4): 250
   sect. *Scandentes* 15(4): 250
  subg. *Nephrangis* 15(4): 249

51

Trillium kamtschaticum 2(1): 229
  smallii 2(1): 230, 230d
Trimeris scaevolifolia 2(4): 377, 378d
Trimezia caerulea 9(1): 9
  connata 1(1): 13
  martinicensis 1(1): 12
  martii 1(1): 11
  sincorana 1(1): 11-14, 12d, 12pl
Tripterospermum japonicum 2(1): 227
Triticum aestivum 2(4): 367
  dicoccoides 2(4): 367
  dicoccon 2(4): 367
Triumfetta semitriloba 14(4): 227
Trochetia melanoxylon 15(1): 29
Trochetiopsis ebenus 15(1): 27-36, 28pl,
  34d
  erythroxylon 2(4): 374; 15(1): 29, 31
  melanoxylon 2(4): 373-374; 15(1): 29
  melanoxylon sensu Cronk 15(1): 33
Trochilidae, Monograph of 11(4): 187, 193
Trollius ledebourii 14(1): 44d
Tropaeolum canariense 9(1): 10, 12
  peregrinum 9(1): 10-12, 10pl
  x *pinnatum* 9(1): 11
Tropical ferns in the Princess of Wales
  conservatory 8(3): 155-158
Tsuga dumosa 13(2): 101
Tulbaghia 8(1): 12
  leucantha 8(1): 12-15, 12pl, 14d
  violacea 8(1): 12
Tulipa sprengeri 15(2): 126
Tupa blanda 10(2): 73
  bridgesii 10(2): 73
  feuillei 5(4): 161
    var. *macrophylla* 5(4): 161
Turkey, Flora of 2(4): 357-368, 359m
  distribution of oaks 2(4): 362m
Turland, Nicholas 8(1): 37-45; 11(3): 117-
  128
Turner, Richard 12(4): 209-220, 211p
Turraea 13(4): 212
Tweedia versicolor 12(4): 237
Tweedie, John 12(4): 237

Two Centuries of Economic Botanists at
  Kew: Part 1 10(2): 84-94; Part 2 10(3):
  132-138
Tye, Alan 15(1): 49-62
Tye, Hilary 15(1): 49-62

Ucriana 9(3): 128, 129
  longifolia 9(3): 128
Ugliest plant in the World, The
  7(2): 85-90
Ulearum sagittatum 12(3): 135-137, 136d,
  136pl
Ulmus wallichiana 5(3): 137-140, 138d
Umbellularia californica 1(3): 133
Urostachys pinifolius 8(3): 127
  piscium 8(3): 127
Urticaceae 14(1): 55
Usambara Violet (*Saintpaulia*): 1(2): 92-93
Uva-ursi glauca 15(3): 166

Vaccinium eberhardtii 1(2): 76
Vallota speciosa 11(1): 10
Vanda hindsii 1(3): 127
Vanderplank, John 13(2): 63-69, 67d;
  15(2): 109-114
van de Kerckhove, Omer 15(1): 43pl
Vanilla albida 13(3): 162
  polylepis 15(3): 169-174, 170pl, 173d
  phalaenopsis 13(3): 162
  sp. 15(3): 173
van Tuyl, Dr J.M. 10(4): 196
Variation in *Fadyenia hookeri* 8(3):
  140-146
Vaughania 14(4): 232, 234, 237
  pseudocompressa 14(4): 237
Vegetative reproduction in pteridophytes
  8(3): 109-112
Veitch, Messrs James & Sons 12(4): 242
Vella pseudocytisus 3(3): 140-142, 141d
  subsp. *pauli* 3(3): 140, 142
  subsp. *pseudocytisus* 3(3): 140, 141
Vellozia sincorana 9(3): 113d
Velloziaceae 9(3): 112

Venus's Fly-trap (*Dionaea muscipula*):
  3(1): 42d
Verbena 13(2): 80
  phlogifolia 12(4): 237
  tweediana 12(4): 237
Verbenaceae 13(2): 79
  subfam. *Verbenoideae* 13(2): 80
Verdcourt, Bernard 3(4): 176-178; 4(3)
  : 141-144; 6(3): 102-106; 13(4): 204-209
Vernonia 13(4): 191
  populifolia 13(4): 192
Viburnum 14(3): 120-124
  bodnantense 14(3): 120
  boninsimense 7(4): 176
  buergeri 7(4): 176
  cotinifolium 13(2): 101
  farreri 14(3): 120
  foetens 14(3): 120-122
  fragrans 14(3): 120
  fusiforme 7(4): 176, 174pl
  grandiflorum 14(3): 120-122
    forma *foetens* 14(3): 120-124
    forma *foetens* 'Desmond Clarke' 14(3):
      120-124, 120pl, 123d
  japonicum 7(4): 173-176, 174pl, 175d
    var. *boninsimense* 7(4): 176
    var. *fusiforme* 7(4): 176
  sect. *Thyrsosma* 14(3): 120
Vicia dennesiana 14(4):
Victoria amazonica 3(4): 168
  regia 3(4): 168; 4(2): 82d
Viguieranthus 14(4): 234
Vireya Rhododendrons 5(3): 99-119

Wahlenbergia mauritiana 7(3): 116
Wakehurst Place 12(1): 28-42
  The Betula collection at 6(2): 65-73
  Recovery from the storm at 6(3): 134-139
Wally's wattle 6(1): 7
Walker, Colin 9(2): 76-85; 11(2): 74-81
Walsh, Wendy 6(3): 110pl; 7(3): 114pl;
  8(1): 12pl, 8(2): 52pl; 10(4): 174pl;
  12(4): 196pl, 198pl

Ward-Hillhorst, E. 10(2): 64pl; 11(4):
  182d, 178pl
Waterlilies: at Oxford 3(4): 167-176
Watson, Mark 10(2): 59-63
Watsonia, A name for the white 11(1): 37
Watsonia alba 11(1): 38-39
  ardernei 11(1): 38-39
  borbonica 11(1): 37-39
    subsp. *ardernei* 11(1): 37-40
    subsp. *borbonica* 11(1): 39
  iridifolia var. *obrienii* 11(1): 38-39
  pyramidata 11(1): 37, 39
  'Ardernei' 11(1): 38-39
  rosea 11(1): 37, 39
  wordsworthiana 11(1): 38-39
  'Arderne's White' 11(1): 38-39
Waxing of a Glorious Rajah, The
  8(2): 81-89
Weberocereus bradei 2(4): 341
  glaber 2(4): 340, 341
  imitans 2(4): 340, 341
  tonduzii 2(4): 339-341, 340pl, 341d
Weinmann, J.A. 14(1): 37
Welwitschia mirabilis 3(1): 46d; 7(2):
  85-90, 88d
  The story of 7(2): 85-90
Werckleocereus glaber 2(4): 341
  imitans 2(4): 341
  tonduzii 2(4): 341
Weyrich, H. 14(1): 40
When is a privet not a lilac? 8(2): 58-63
White, Frank 6(4): 166-171
Wickens, G.E. 10(2): 84-94; 10(3): 132-
  138
Wickison, Sue 4(2): 60pl
Wijnands, D. Onno 4(2): 78-91
Wilcoxia kroenleinii 2(2): 265
  poselgeri 2(2): 265
  tamaulipensis 2(2): 263
  tuberosa 2(2): 265
Wildlife and Countryside Act 1(1): 25
Wilford, R. 12(3): 168-173; 15(3): 180-
  185; 15(4): 220-225, 226-230

Wilkin, Paul 13(1): 22-26; 14(1): 7-12
Willkomm, H.M. 14(1): 37, 41-43
Windmill plant 13(1): 19-20
Winter Aconite 4(1): 35-36
Wise, Rosemary 6(4): 166-171, 170pl
Wisteria 14(4): 178, 182
Wollaston, William Hyde 15(4): 266
Wong, K.M. 14(2): 82-86
Wood, Jeffrey 1(3): 99, 103-105, 139-42; 2(1): 211-213, 2(4): 345-348; 3(1): 14-38; 3(4): 147-151; 4(2): 62-64, 74-77; 5(4): 154-156; 9(2): 55-59; 10(1): 9-12, 18-22, 36-38, 10(4): 161; 11(1): 22-28; 13(3): 139-145
Woodgyer, Elizabeth 12(4): 186-190
Woodin, Carol 12(2): 64pl; 13(3): 130pl
Woods, Patrick 1(1): 2, 9-11; 2(3): 291-308; 8(1): 15-23
World Conservation Strategy 1(1): 23
World Wildlife Fund 1(2): 92-93; 1(3): 98
Woy, Urs 9(2): 60pl
Wyk, Abraham E. van 14(1): 16-22
Wyse-Jackson, Peter S. 7(3): 113-117

Xeranthemum 13(2): 57
Xerophyta plicata 9(3): 107
Ximenez, R. 13(3): 132d
Xiphion aitchisonii 8(4): 174
  alatum 3(2): 65
Xiphium planifolium 3(2): 65

Yeo, Peter 1(2): 99, 111-118
Yong Shik Kim 15(2): 141-146

Zaluzianskya maritima
  var. *atro-purpurea* 5(2): 72
  var. *breviflora* 5(2): 72
  var. *grandiflora* 5(2): 72
  microsiphon 5(2): 69-72
Zappi, Daniela 9(3): 134-138; 10(3):111-113, 1113d, 10(4): 166-170; 13(2): 70-78; 14(1): 13-15; 14(3): 120-124
Zephyranthes aurea 6(3): 101

boliviensis 6(3): 98
flammea 6(3): 101
flava 6(3): 101
tubiflora 6(3): 101
  var. *flammea* 6(3): 101
Zeuxine ballii 8(2): 79d, 80
Zingiber gracile
  var. *aurantiacum* 12(2): 73-75
  griffithii 12(2): 73-75
  sulphureum 12(2): 73-77, 74pl, 76d
Zingiberaceae 11(1): 13; 11(3): 104; 12(2): 73-77; 13(1): 8-13; 14(1): 2; 14(2): 77; 15(4): 220, 226
Zollernia 14(4): 194-197
  splendens 14(4): 194-197, 194pl, 196d
  tango 14(4): 194
Zygia 14(4): 187, 243

CURTIS'S BOTANICAL MAGAZINE
A-Z list of species illustrated in Series 6, Volumes 1-18 (1984-2001)
[Showing volume number, plate number and year]

Abarema idiopoda	14	326	1997	Arctotis 'Nicholas Hind'	9	207	1992
Abies recurvata	16	369	1999	Arisaema bottae	4	76	1987
Acacia chamelensis	10	219	1993	Arisaema consanguineum	1	9	1984
Acacia pataczekii	6	116	1989	Arisaema costatum	1	7	1984
Acanthostachys strobilacea	8	173	1991	Arisaema exappendiculatum	4	81	1987
Acanthus dioscoridis	7	157	1990	Arisaema filiforme	2	46	1985
Adenophylla triphylla var. japonica	16	365	1999	Arisaema kiushianum	16	364	1999
Aechmea abbreviata	1	2	1984	Arisaema leschenaultii	1	10	1984
Aechmea streptocalycoides	2	26	1985	Arisaema sikokianum	3	64	1986
Aerangis arachnopus	11	242	1994	Arisaema tortuosum	1	8	1984
Aerangis distincta	13	300	1996	Arisaema undulatifolium	1	11	1984
Aerangis spiculata	4	77	1987	Arisaema yamatense ssp. sugimotoi	1	12	1984
Aerangis verdickii	15	333	1998	Arisarum proboscideum	7	142	1990
Aeschynanthus angustifolius	8	168	1991	Arisarum simorrhinum	7	143	1990
Aeschynanthus chrysanthus	8	167	1991	Arisarum vulgare	7	144	1990
Albuca pendula	14	312	1997	Arrojadoa bahiensis	13	291	1996
Allium schmitzii	11	235	1994	Arrojadoa penicillata	13	292	1996
Allium virgunculae	6	124	1989	Arum pictum	5	102	1988
Alocasia alba	17	403	2000	Asarum caulescens	11	243	1994
Alocasia melo	14	315	1997	Asarum fudsinoi	9	197	1992
Alocasia nebula	17	381	2000	Asarum minamitanianum	9	198	1992
Aloe rauhii	3	57	1986	Asarum yakusimense	11	244	1994
Amorphophallus kiusianus	17	380	2000	Augusta longifolia	9	202	1992
Ampelopsis stylosa	6	123	1989	Azara eryngioides	7	154	1990
Amydrium zippelianum	12	269	1995	Barkeria melanocaulon	5	111	1988
Andira inermis subsp. inermis	17	399	2000	Barringtonia edulis	9	210	1992
Androsiphon capense	7	155	1990	Barringtonia novae-hiberniae	9	208	1992
Angiopteris elliptica	8	179	1991	Barringtonia procera	9	209	1992
Ansellia africana	10	216	1993	Bauer colour chart	14	317	1997
Anthurium warocqueanum	12	271	1995	Bauhinia jenningsii	11	254	1994
Antirrhinum grosii	6	129	1989	Begonia amphioxus	7	149	1990
Arachnis longisepala	9	194	1992	Begonia bogneri	3	65	1986
Araucaria bidwillii	8	185	1991	Begonia cauliflora	7	147	1990
Arctostaphylos glauca	15	344	1998	Begonia erythrogyna	7	150	1990

Begonia goudotii	18	407	2001
Begonia imbricata	7	146	1990
Begonia kinabaluensis	7	148	1990
Begonia malachosticta	7	145	1990
Bellevalia hyacinthoides	11	252	1994
Betula chinensis	6	126	1989
Betula utilis var. jacquemontii 'Grayswood Ghost'	6	125	1989
Biarum ditschianum	12	275	1995
Bossiaea walkeri	11	255	1994
Bouvardia laevis	10	232	1993
Bouvardia ternifolia	10	233	1993
Brighamia citrina var. napaliensis	3	60	1986
Bucephalandra motleyana	12	272	1995
Bursaria spinosa	18	425	2001
Calanthe izuinsularis	18	417	2001
Calanthe striata	18	418	2001
Calanthe tricarinata	18	416	2001
Callicarpa japonica	16	363	1999
Calypso bulbosa var. occidentalis	3	67T	1986
Calypso bulbosa var. speciosa	3	67B	1986
Camellia flava	18	426	2001
Campanula betulifolia	17	389	2000
Camptosema scarlatinum var. pubescens	9	190	1992
Campylotropis polyantha	14	327	1997
Centropogon cornutus	9	187	1992
Cephalipterum drummondii	16	374	1999
Ceropegia craibii	18	430	2001
Chamaedorea metallica	4	87	1987
Choisya dumosa subsp. arizonica	15	346	1998
Christia vespertilionis	13	286	1996
Cirsium subcoriaceum	6	137	1989
Cleistocactus acanthurus & sextonianus	15	338	1998
Clematis confusa	8	181	1991
Clematis marmoraria	4	82	1987
Clowesia rosea	7	140	1990
Coccocypselum guianense	10	221	1993
Cocos nucifera	16	355	1999

Cornus 'Porlock'	7	162	199
Corydalis aitchisonii	16	378	199
Corydalis flexuosa	15	332	199
Corydalis popovii	5	100	198
Cotoneaster bradyi	12	280	199
Crassula tetragona subsp. acutifolia	18	409	200
Crocus kerndorffiorum	15	342	199
Crocus paschei	18	410	200
Crotalaria quinquefolia	14	328	199
Cryptomeria japonica	16	371	199
Culcasia seretii	12	274	199
Cuminia eriantha	3	68	198
Cupressus cashmeriana	11	253	199
Cyclamen colchicum	15	347	199
Cyclamen purpurascens	3	63	198
Cylindrocline commersonii	13	303	199
Cymbidium kanran	16	366	199
Cynorkis gibbosa	13	296	199
Cyperus prolifer	11	236	199
Cypripedium henryi	14	324	1997
Cypripedium japonicum var. formosanum	4	74	1987
Cyrtanthus herrei	11	237	1994
Daphne jasminea	11	251	1994
Daphne kiusiana	6	131	1989
Daphniphyllum himalaense subsp. macropodum	16	376	1999
Dendranthema yezoense	5	98	1988
Dendrobium alexandrae	7	153	1990
Dendrobium cuthbertsonii	2	37	1985
Dendrobium eximium	3	56	1986
Dendrobium gouldii	10	213	1993
Dendrobium hellwingianum	2	38	1985
Dendrobium rarum	2	30	1985
Dendrobium rhodostictum	2	41	1985
Dendrobium tapiniense	2	42	1985
Dendrobium vexillarius	2	39	1985
Dendrobium violaceum	2	40	1985
Deutzia purpurascens 'Alpine Magician'	10	234	1993

iascia rigescens	3	61	1986	Eryngium leavenworthii	16	356	1999
illenia alata	14	316	1997	Erythronium japonicum	18	411	2001
imorphanthera kempteriana	1	3	1984	Espeletia schultzii	7	158	1990
ionysia denticulata	5	92	1988	Eucomis schijffii	6	121	1989
ionysia involucrata	2	28	1985	Eulophia pulchra	13	294	1996
iplazium proliferum	8	178	1991	Euphorbia schillingii	4	80	1987
isa cardinalis	5	94R	1988	Eurychone galeandrae	18	423	2001
isa 'Kirstenbosch Pride'	5	94L	1988	Eurychone rothschildiana	18	424	2001
istephanus populifolius	13	302	1996	Exochorda giraldii	3	69	1986
odecatheon hendersonii	14	322	1997	Fadyenia hookeri	8	180	1991
racula vampira	13	295	1996	Falcatifolium taxoides	16	370	1999
uperrea pavettifolia	15	336	1998	Fargesia murieliae	8	186	1991
chinocereus adustus ssp. schwarzii	2	35	1985	Fitzroya cupressoides	16	372	1999
chinocereus chisoensis fobeanus	2	33	1985	Fritillaria affinis	17	395	2000
chinocereus dasyacanthus	1	24	1984	Fritillaria chitralensis	13	288	1996
chinocereus engelmannii engelmannii	2	31B	1985	Fritillaria crassifolia subsp. kurdica	17	396	2000
chinocereus fasciculatus ssp. fasciculatus	2	31T	1985	Fritillaria davidii	17	382	2000
chinocereus polyacanthus ssp. acifer	1	20	1984	Fritillaria delavayi	9	193	1992
chinocereus poselgeri	2	34	1985	Fritillaria involucrata	17	398	2000
chinocereus pulchellus ssp. pulchellus	2	36B	1985	Fritillaria pudica	17	391	2000
chinocereus pulchellus ssp. weinbergii	2	36T	1985	Fritillaria stenanthera	17	398	2000
chinocereus rigidissimus ssp. rubispinus	1	24	1984	Fritillaria thunbergii	17	393	2000
chinocereus russanthus	1	23	1984	Fritillaria tuntasia	17	397	2000
chinocereus scheerii ssp. gentryi	1	19	1984	Fritillaria uva-vulpis	17	392	2000
chinocereus stoloniferus ssp. stoloniferus	2	32	1985	Gasteria baylissiana	16	359	1999
chinocereus stramineus	1	21	1984	Geranium cataractarum	8	169	1991
chinocereus subinermis	11	246	1994	Geranium orientalitibeticum	1	18	1984
chinocereus viereckii	1	22	1984	Geranium pylzowianum	1	17	1984
laphoglossum crinitum	8	176	1991	Glaucium flavum	6	133	1989
ncyclia mariae	2	25	1985	Gloxinia sylvatica	12	278	1995
pidendrum ilense	8	164	1991	Guzmania multiflora	8	172	1991
pigeneium triflorum	13	299	1996	Gynura scandens	7	151	1990
pimedium flavum	12	263	1995	Habranthus martinezii	4	79	1987
pipactis royleana	10	214	1993	Harveya pulchra	12	281	1995
rigeron oharai	4	83	1987	Helichrysum meyerijohannis	12	259	1995
riobotrya japonica	18	419	2001	Helleborus thibetanus	15	353	1998
				Helleborus X nigercors	4	85	1987

Heteropolygonatum ogisui	18	415	2001
Hexastylis speciosa	5	110	1988
Hibiscus fragilis	13	306	1996
Hibiscus kokio subsp. saintjohnianus	15	352	1998
Himantoglossum caprinum	3	52	1986
Himantoglossum hircinum	11	240	1994
Hippeastrum puniceum	14	318	1997
Holmskioldia sanguinea	13	293	1996
Hoya multiflora	7	139	1990
Huernia kennedyana	15	331L	1998
Huernia macrocarpa	15	331U	1998
Huperzia pinifolia	8	177	1991
Hyophorbe lagenicaulis	13	301	1996
Hypericum buckleyi	11	245	1994
Hypericum subsessile	15	337	1998
Ilex cyrtura	1	13	1984
Impatiens columbaria	15	335	1998
Impatiens flanaganae	2	44	1985
Impatiens kilimanjari (ssp. pocsii & ssp. kilimanjari)	14	311	1997
Indigofera jucunda	14	329	1997
Iris aitchisonii	8	184	1991
Iris dolichosiphon	7	141	1990
Iris lazica	16	357	1999
Iris planifolia	3	58	1986
Iris sprengeri	17	388	2000
Iris tenax	8	165	1991
Jasminum leptophyllum	17	384	2000
Jasminum officinale 'Inverleith'	9	196	1992
Jasminum sinense	10	224	1993
Jumellia walleri	17	402	2000
Kleinia saginata	6	134	1989
Lachenalia elegans var. flava	18	408	2001
Lachenalia violacea	16	373	1999
Laelia bahiensis	9	199	1992
Lecanopteris lomarioides	12	270	1995
Lemboglossum cervantesii	5	97	1988
Leontochir ovallei	14	308	1997

Leptospermum lanigerum	12	277	19
Leucojum tingitanum	9	206	19
Ligustrum sempervirens	8	171	19
Lilium medeoloides	18	412	20
Limodorum abortivum	10	230	19
Limodorum abortivum	10	231	19
Lindheimera texana	7	156	199
Liquidambar orientalis	17	386	20
Lobelia bridgesii	10	220	199
Lobelia organensis	9	200	199
Lobelia tupa	5	112	198
Lobelia vagans	13	304	199
Lonicera pilosa	10	223	199
Loropetalum chinense	18	421	200
Lycaste candida	10	215	199
Lycaste tricolor	1	14	198
Lycoris sanguinea	16	361	199
Magnolia 'Albatross'	2	27	198
Mahonia confusa	10	229	199
Mahonia russellii	6	122	198
Maingaya malayana	10	222	199
Malaxis punctata	4	78	198
Mandevilla illustris	9	203	199
Marcgravia umbellata	14	321	199
Marshallia grandiflora	15	343	1998
Medusagyne oppositifolia	6	138	1989
Melianthus villosus	4	84	1987
Mexipedium zerophyticum	13	297	1996
Micromeria marginata	18	420	2001
Mimetes chrysantha	11	257	1994
Mimulus naiandinus	17	400	2000
Modiolastrum gilliesii	18	431	2001
Mossia intervallaris	14	310	1997
Neoglaziovia variegata	9	201	1992
Neomarica caerulea	9	188	1992
Nephrolepis cordifolia	8	175	1991
Nesocodon mauritianus	7	152	1990

Nomocharis pardanthina	11	239	1994	Passiflora cuneata 'Miguel Molinari'	15	340	1998
Ochagavia elegans	13	287	1996	Passiflora mollissima	11	258	1994
Odontoglossum harryanum	13	298	1996	Passiflora quadrifaria	13	290	1996
Oeceoclades decaryana	18	405	2001	Pelargonium endlicherianum	10	227	1993
Ophrys argolica	3	49L	1986	Pelargonium nephrophyllum	18	422	2001
Ophrys doerfleri	3	50R	1986	Pelargonium quercetorum	10	228	1993
Ophrys lutea	3	50L	1986	Phaeonocoma prolifera	13	289	1996
Ophrys oestrifera var. heldreichii	3	49R	1986	Phaius pulchellus var. sandrangatensis	16	375	1999
Ophrys spruneri	2	45	1985	Philodendron billietiae	13	285	1996
Orchis anatolica	3	53	1986	Photinia niitakayamensis	5	109	1988
Orchis boryi	3	54R	1986	Phragmipedium besseae	6	135	1989
Orchis sancta	10	212	1993	Phyllostachys aureosulcata	12	261	1995
Orchis simia	3	54L	1986	Phyllostachys dulcis	12	262	1995
Origanum amanum	11	249	1994	Phyllostachys viridiglaucescens	12	260	1995
Origanum dictamnus	11	248	1994	Pilea peperomiodes	1	5	1984
Origanum rotundifolium	11	250	1994	Pinellia cordata	18	429	2001
Orthophytum albopictum	2	48	1985	Pinellia tripartita	5	95	1988
Otacanthus coeruleus	8	182	1991	Pinus brutia	16	367	1999
Oxalis hirta	10	217	1993	Platycladus orientalis	16	368	1999
Pachpodium rosulatum	7	159	1990	Plectocephalus rothrockii	13	283	1996
Paeonia japonica	18	413	2001	Pleione aurita	5	113	1988
Paeonia turcica	17	390	2000	Pleione scopulorum	5	114	1988
Papaver fauriei	16	362	1999	Polygonatum cyrtonema	16	358	1999
Papaver lateritium	17	387	2000	Polystachya melliodora	5	96	1988
Paphiopedilum bougainvilleanum	3	71	1986	Polyxena corymbosa	5	91	1988
Paphiopedilum exul	10	211	1993	Protea aurea var. aurea	11	256	1994
Paphiopedilum fairrieanum	2	47	1985	Pseudogynoxys cabrerae	9	205	1992
Paphiopedilum hennisianum	3	59	1986	Pseudophegopteris levingei	12	279	1995
Paphiopedilum malipoense	4	86	1987	Pycnospatha arietina	10	226	1993
Paphiopedilum sanderianum	1	1	1984	Pyrolirion tubiflorum	6	127	1989
Paphiopedilum schoseri	12	265	1995	Quesnelia humilis	8	174	1991
Paphiopedilum urbanianum	3	62	1986	Ramosmania rodriguesii	13	305	1996
Paphiopedilum violascens	3	70	1986	Ranunculus weberbaueri	15	351	1998
Paphiopedilum wentworthianum	3	72	1986	Rhaphidophora glauca	16	377	1999
Papuacedrus papuana	12	266	1995	Rhipsalis pilocarpa	14	320	1997
Paramongaia weberbaueri	14	323	1997	Rhipsalis puniceodiscus	16	360	1999
Parochetus africanus	8	170	1991	Rhododendron ?javanicum ssp. brookeanum	9	192	1992

Rhododendron burtii	6	130	1989
Rhododendron christii	2	29	1985
Rhododendron culminicolum v. angiense	12	268	1995
Rhododendron dalhousiae v. dalhousiae	4	73	1987
Rhododendron dielsianum	1	16	1984
Rhododendron edgeworthii x ciliatum 'Princess Alice'	9	191	1992
Rhododendron konori	5	107	1988
Rhododendron lineare	5	103	1988
Rhododendron pleianthum	4	88	1987
Rhododendron praetervisum	5	104	1988
Rhododendron rarilepidotum	1	6	1984
Rhododendron russatum	7	161	1990
Rhododendron salicifolium	5	106	1988
Rhododendron ssp. discolor	3	55	1986
Rhododendron stenophyllum	5	105	1988
Rhododendron superbum	5	108	1988
Rhus batophylla	10	218	1993
Rhynchospora nervosa	10	225	1993
Romulea hirta	1	15	1984
Rosa complicata	14	313	1997
Rosa cymosa 'Rebecca Rushforth'	14	309	1997
Rosa glauca	6	115	1989
Roscoea alpina	14	314	1997
Roscoea capitata	15	350	1998
Roscoea ganeshensis	13	284	1996
Roscoea humeana f. lutea	17	383	2000
Roscoea praecox	14	307	1997
Roscoea purpurea 'Red Gurkha'	11	247	1994
Roscoea schneideriana	11	238	1994
Roscoea tumjensis	15	349	1998
Rosularia muratdaghensis	9	195	1992
Rubus rosifolius 'Coronarius'	15	341	1998
Salix koriyanagi	12	282	1995
Salvia albimaculata	17	385	2000
Salvia darcyi	11	241	1994
Salvia lanceolata	8	183	1991
Sarracenia psittacina	5	99	1988

Scilla maderensis	5	93	19
Scutellaria longituba	18	414	20
Scyphochlamys revoluta	6	128	19
Sedum lucidum	4	75	19
Senecio formosus	8	163	19
Serapias lingua	3	51L	19
Serapias vomeracea var. orientalis	3	51R	19
Sisyrinchium palmifolium	15	339	19
Skimmia japonica 'Chilan Choice'	4	89	19
Skimmia japonica 'Wakehurst White'	12	264	19
Skimmia laureola var. multinervia	4	90	198
Skimmia x confusa 'Chelsea Physic'	6	136	198
Smilax megalantha	3	66	198
Solanum uleanum	18	427	200
Sophora davidii 'Hans Fliegner'	17	379	200
Sophora toromiro	14	330	199
Spigelia flava	9	204	199
Stenoglottis fimbriata	6	117	198
Stenoglottis longifolia	6	120	198
Stenoglottis woodii	6	118	198
Stenoglottis zambesiaca	6	119	198
Steudnera discolor	12	276	199
Strophanthus sarmentosus	17	401	200
Syringa protolaciniata 'Kabul'	6	132	198
Syringa x laciniata	6	132	198
Toona sinensis	15	348	198
Tridactyle truncatiloba	15	354	198
Trimezia sincorana	1	4	198
Trochetiopsis ebenus	15	334	198
Tropaeolum peregrinum	9	189	1992
Tulbaghia leucantha	8	166	1991
Tulipa regelii	18	406	2001
Ulearum sagittatum	12	273	1995
Vanilla polylepis	15	345	1998
Veronicastrum stenostachyum	18	428	2001
Viburnum grandiflorum foetens 'Desmond Clarke'	14	319	1997
Viburnum japonicum	7	160	1990

Viola hederacea	17	404	2000
Weberocereus tonduzii	2	43	1985
Zaluzianskya microsiphon	5	101	1988
Zingiber sulphureum	12	267	1995
Zollernia splendens	14	325	1997